Palgrave Macmillan Studies in Banking and Financial Institutions

Series Editor
Philip Molyneux, Bangor University, Bangor, UK

The Palgrave Macmillan Studies in Banking and Financial Institutions series is international in orientation and includes studies of banking systems in particular countries or regions as well as contemporary themes such as Islamic Banking, Financial Exclusion, Mergers and Acquisitions, Risk Management, and IT in Banking. The books focus on research and practice and include up to date and innovative studies that cover issues which impact banking systems globally.

James Atta Peprah · Evelyn Derera ·
Harold Ngalawa · Thankom Arun
Editors

Financial Sector Development in Ghana

Exploring Bank Stability, Financing Models, and Development Challenges for Sustainable Financial Markets

palgrave
macmillan

Editors
James Atta Peprah
Department of Applied Economics
University of Cape Coast
Cape Coast, Ghana

Harold Ngalawa
College of Law and Management Studies
University of KwaZulu-Natal
Durban, South Africa

Evelyn Derera
College of Law and Management Studies
University of KwaZulu-Natal
Pietermaritzburg, South Africa

Thankom Arun
Essex Business School
University of Essex
Colchester, UK

ISSN 2523-336X ISSN 2523-3378 (electronic)
Palgrave Macmillan Studies in Banking and Financial Institutions
ISBN 978-3-031-09344-9 ISBN 978-3-031-09345-6 (eBook)
https://doi.org/10.1007/978-3-031-09345-6

This Palgrave Macmillan imprint is published by the registered company Springer Nature Switzerland AG
The registered company address is: Gewerbestrasse 11, 6330 Cham, Switzerland

Acknowledgements

The editors acknowledge the scholarly contributions of the authors for each of the chapters. We sincerely appreciate the generous time and effort they put into the research and writing of these chapters. We also sincerely appreciate the contributions of anonymous referees for their invaluable and constructive feedback that has helped to enhance the quality of this book. This collection of edited chapters on Financial Sector Development in Ghana would not have been possible without the selfless support of Palgrave Macmillan Publishers. We acknowledge their support with appreciation. Finally, we thank the Universities of Cape Coast in Ghana, KwaZulu-Natal in South Africa and Essex in the UK, to which the editors are affiliated.

Contents

Notes on Contributors

Abe Omo Oyeniyi is a law Lecturer at the Huddersfield Business School, University of Huddersfield, Queensgate Huddersfield, UK. He specialises in teaching commercial, indigenous dispute resolution, contract law and practised as a corporate and commercial law Attorney of Nigeria for several years. He had previously taught at the Universities of Pretoria and Cape Town, both in South Africa, and Afe Babalola University, Nigeria. In 2015/2016, he was a Fulbright Scholar at Loyola University, Chicago, USA. He is currently also serving as a Research Associate at the Centre for Comparative Law in Africa, University of Cape Town, South Africa. He has spoken at local and international conferences on various areas of his research interest and has published in peer-reviewed journals. He serves on the Executive Council of the International Law Association, Nigerian Branch and a member of the IUCN World Commission on Environmental Law, Nigerian Bar Association, Global Business & Human Rights Scholars Association, and the Business & Human Rights Teaching Forum Meeting.

Adjei Prince is an Economist with the Department of Economics, University of Ghana. He is an experienced academic and researcher whose keen focus is on Economics of Education and Labour markets. His prime research goal is to study the forward and backward linkages between school to work transitions and how institutions play a role in ensuring harmony in the development of skills. As a development economist, his research cuts across several fields including poverty and inequality, applied microeconomics and economic policy. He holds a Ph.D. in Development Economics, M.Phil. and Bachelor's degree in Economics all from the University of Ghana.

Afful Benedict Jr. is a Lecturer in Economics, obtained his Ph.D. and M.Phil. in Economics from the University of Cape Coast where he has considerably taught and research. He also has a Diploma in Project Management from Institute of Commercial Management UK, certificate in Research Ethics Training Curriculum, *fhi 360*, USA, and certificate in Cost Analysis in Education, UNESCO and International Institute for Educational Planning. In the past, he has worked with Ghana Statistical Service, Electoral Commission of Ghana, IFPRI, ILO, UNICEF, University of Development Studies and Department of Economics, University of Cape Coast.

Akande Joseph Olorunfemi is an Associate Professor in Accounting Science with Walter Sisulu University, South Africa and a member of multidisciplinary research working groups. His research interest is in the broad area of finance and accounting with expertise in panel data and time series analysis.

Annim Samuel Kobina is a Professor of Economics with specific concentration on Micro Development Economics and applied Micro econometrics at the Department of Economics, University of Cape Coast, Ghana. He assumed office as the Government statistician at the Ghana Statistical Services on March 1, 2019. He earned Bachelor of Arts, Master of Philosophy and Doctor of Philosophy degrees in Economics at the University of Cape Coast (Ghana), University of Ghana (Ghana) and University of Manchester (United Kingdom), respectively. Following the completion of his doctoral studies, he was engaged at the University of Manchester and University of Lancashire, both in the UK, in the respective capacities of Research Associate and Post-doctoral Research Fellow. He has approximately 20 years of teaching experience in Universities both in Ghana and abroad, and has more than 40 peer-reviewed journal articles, book chapters and technical reports to his credit. His publications are available in academic outlets such as *World Development*, *Journal of Development Studies*, *Journal of International Development* and the Oxford University Press. Samuel passionately supports national and global development agenda by providing professional service to the National

Development Planning Commission, the National Statistical System in Ghana and several international bodies.

Arun Shoba is a Professor of Applied International Sociology at the Manchester Metropolitan University. She has many years of experience in teaching in sociology and international development. Her research goals are to better understand processes of global social change, as these processes are expressed in particular social and spatial contexts and differently, among diverse social constituencies. Her scholarship on sociology of development is seen in her overall research, teaching and outreach activities, particularly on the causes, dynamics and consequences of social and economic change, and engages this approach to undertake a number of parallel but distinct research interests. Foremost, her research publications and expertise concern gender matters in the global society and the knowledge economy. This includes research in the areas of neo-liberal policies and impact on digital technologies, and its impact on gender and the labour market. In doing so, she has an international standing in this field of gender research in a sociological analysis of development informatics. Her publications have informed policy reports and scholarly debates on the impact of IT-based services within the gendered labour market, thus building such knowledge through engaged research with a wide range of constituencies. Her research into global mobilities among skilled migrants also exposes gendered and racialised processes of global change.

Arun Thankom is a Professor of Global Development and Accountability at the Essex Business School. Currently, he is a Professor Extraordinaire at the Stellenbosch Business School, South Africa and a Research Fellow at IZA, Bonn. He is also Chairing an academic steering group on Financial inclusion in the International Cooperative and Mutual Insurance Federation (ICMIF). He is a Fellow of the Academy of Social Sciences and a Fellow of the Royal Society of Arts, Manufactures and Commerce. He works towards an interdisciplinary learning process to understand the global challenges from a Development finance perspective. Over the years, the research aims to understand, theorise and tackle the problems created by the uneven relationships between business, society and the economy in an interdisciplinary framework. His recent research interests are Fintech, financial inclusion, climate change and sustainability.

Ayisi Richard K. is an Economist with the Department of Economics of University of Ghana. He has over decade-long experience in economic policy, providing technical support in the areas of research and policy advice to both private enterprises, government and international organisation. His research focusses on incorporating the idiosyncratic characteristics of developing economics in economic models to develop policies that incorporate country's specific characteristics. His research interest spans across development macroeconomics, monetary and financial economics, regulatory economics, poverty and inequalities, and adaptation strategies. He holds a Ph.D. Economics degree from University of Milan, and M.Phil. and bachelor's degree from University of Ghana.

Amoo Bondzie Eric is currently a Lecturer at the Department of Economic Studies, University of Cape Coast, Ghana where he teaches both at the undergraduate and graduate levels. He holds a master's degree in Economic Policy and Institutions from La Sapienza University of Rome and a Ph.D. in Economics from the Catholic University of Milan all in Italy. His research areas are Macroeconomics, Monetary Economics, DSGE modelling and Time Series Econometrics.

Bose Udichibarna is a Senior Lecturer (Associate Professor) in Finance at the Essex Business School, University of Essex. She is a Fellow of the Higher Education Academy (FHEA) and her research interest lies in the areas of financial economics, corporate finance, financial inclusion, sustainable finance and trade. She is currently on the editorial board of several journals including the *International Journal of Finance & Economics* and *The Journal of Risk Finance*. She obtained her Ph.D. in Economics from the University of Glasgow and was awarded the Adam Smith Business School (ASBS) Prize for Ph.D. Excellence. Prior to that, she earned her Master's degree in Economics from the University of Essex and Bachelor's degree in Economics from the University of Delhi. She has received several research grants from the Economic and Social Research Council (ESRC), the UK–India Education and Research Initiative (UKIERI) and the Global Challenges Research Fund (GCRF).

Cantah William Godfred holds a Ph.D. in Economics from the University of Cape Coast with specific research interest in Monetary Policy, Finance and International Economics. He is currently a Lecturer with the Department of Data Science and Economic Policy at the University of Cape Coast. He has consulted for a number of organisations including the Ghana Statistical Service and the African Development Bank.

Darfor Kwabena Nkansah is currently affiliated to the University of Cape Coast as a Senior Lecturer in Economics at the Department of Applied Economics, School of Economics, University of Cape Coast. He has expertise in Natural Resources and Environment Economic Issues, Labour Issues, Economic and Biophysical impacts of Climate Change, Business Simulation, Finance and Forecasting.

Derera Evelyn is an Associate Professor of Entrepreneurship in the School of Management, IT & Governance at the University of KwaZulu-Natal (UKZN). She holds a Ph.D. in Entrepreneurship from UKZN. She is passionate about entrepreneurship, particularly the role of women entrepreneurs in developing nations. She also holds a Diploma in Banking from the Institute of Bankers of Zimbabwe. Before joining the academia, she worked in the banking sector in Zimbabwe where she held several positions including Customer Relationship Manager in Private Banking. Her financial sector experience spans over ten years.

Ewusie Ewura-Adwoa is a Lecturer at the School of Economics, University of Cape Coast, Ghana. She holds a B.A. in Economics from the University of Cape Coast, an M.Phil. in Development Studies from the University of Cambridge, UK and a Ph.D. in Economics from the University of Cape Coast. Her background is in Development Economics, Institutions and Development, Globalisation and Policy Evaluation. Her research interests embody issues of poverty and inequality; female economic empowerment and domestic violence; financial inclusion, health and education. She has previously worked in Local Government, Community Economic and Social Development, and Business Development (UK and Ghana).

Gyimah Kofi Nyarko is a Ph.D. candidate in the School of Accounting, Economics and Finance, University of KwaZulu-Natal, South Africa. He has an M.B.A. in Finance from the University of Leicester, UK. His research interest is in entrepreneurial finance. He is currently a full-time Lecturer in Banking and the programme manager of Banking and Finance Department at Pentecost University.

Korankye-Sakyi Francis Kofi is a Lecturer at the Faculty of Law, University of Cape Coast, Ghana. He specialises in teaching International Trade and Investment, Local Government, Commercial and Company Law. He is the Founder and a Senior Private Legal Consultant at Centre for African Trade and Investment Policies, Ghana. He is National Council

Member for the Ghana National Chamber of Commerce and Industry. He is a former District Chief Executive for the Twifo Hemang Lower Denkyira District in the Central, region, Ghana from 2013 to 2017. In 2018, he received an ABSA Bank scholarship through the International Development Law Unit of the Centre for Human Rights, Faculty of Law, University of Pretoria to read for his LL.M. (with Thesis) in International Trade and Investment Law in Africa. He also holds Advanced Certificates in Civil Society Law in Africa; Business and Human Rights Law in Africa; Rights to Development; Results-Based Monitoring and Evaluation in Development Projects and has attended many conferences and workshops relevant to his specialised areas. He has a number of researched articles and publications to his credit.

Muzindutsi Paul-Francois is a Professor of Finance in the School of Accounting, Economics and Finance, University of KwaZulu-Natal. His research interests include financial markets, behavioural finance, financial risk management, time series analysis of macroeconomic variables and development economics.

Ngalawa Harold is a Professor of Economics in the School of Accounting, Economics and Finance at the University of KwaZulu-Natal (UKZN) in South Africa. In 2018, he founded the Macroeconomics Research Unit at UKZN, the only research unit in South Africa specialising in macroeconomics research. He has also taught Economics at the University of Malawi. Before joining the academia, he worked as a Bank Economist at the Commercial Bank of Malawi Limited and later as Head of Research at Continental Discount House Limited in Malawi. He holds a Ph.D. in economics from the University of Cape Town and his research focus is on monetary theory and practice, deposit insurance, bank instability, structural asymmetries and policy failure.

Nunoo Jacob is a Senior Lecturer at the Department of Applied Economics at the University of Cape Coast, Ghana. He holds a Master of Philosophy degree in Economics. He is currently pursuing a Ph.D. in Economics at the University of Cape Coast. He is a versatile Lecturer who is ready to collaborate on diverse academic disciplines. His research interest is in the field of labour economics and economics of education. He has a number of peer-reviewed articles and a book chapter to his credit.

Nyebar Abraham is a Finance graduate in the School of Accounting, Economics and Finance at the University of KwaZulu-Natal. He is a qualified Chartered Accountant (CA) and a member of Chartered Institute of Supply Chain Management (CISCM). He currently works as a Chief Finance Officer at Pentecost University, Accra, Ghana. His research interests include Finance, Auditing, Tax and Procurement.

Obalade Adefemi A. is a Post-doctoral Research Fellow (Finance) in the School of Accounting, Economics and Finance at the University of KwaZulu-Natal. His research interests include financial market, financial economics and time series analysis.

Obeng Camara Kwasi is a Senior Lecturer in the Department of Economic Studies, School of Economics, University of Cape Coast. He obtained his Ph.D. in Economics from the University of Cape Coast in 2011. He was a Global Academic Partnership (Gap)Scholar at the University of South Florida in 2010 and an African Economic Research Consortium (AERC)/IMF visiting Scholar to the IMF in 2007. He has published extensively and attended many conferences and he is an active member of the AERC, among many others. His research interests are international trade and finance, fiscal implications of trade policy, and lately, machine learning.

Ofori Isaac Kwesi is a Ph.D. Student in the Department of Economics, University of Insubria, Varese, Italy. He holds M.Phil. and B.A. degrees in Economics both from the University of Cape Coast, Ghana. He was formerly a Research Assistant at the Department of Data Science and Economic Policy, University of Cape Coast, Ghana. His research interest is in development economics with focus on cross-country inequality, machine learning, and microsimulation. He has teamed up with a number of consultants to implement research projects in agriculture, health and environmental protection for institutions like USAID, ACDI/VOCA, Catholic Relief Services and UNFPA.

Peprah James Atta is Associate Professor in Economics at the Department of Applied Economics at the School of Economics, University of Cape Coast, Ghana. He holds Ph.D. in Economics, M.Phil. in Economics and Bilingual M.B.A. (Banking and Finance). He is part-qualified Accountant (Institute of Chartered Accountants-Ghana). He is currently the Dean of School of Economics at the University of Cape Coast. His areas of research include development economics, financial markets and institutions in developing countries, microfinance, economic growth and development as well as Small and Medium Scale Enterprise (SMEs) development. His current research work focusses on energy financing, digitised financial services and access to finance by people living with disability. He has a number of publications that focus on SMEs, microfinance and poverty in Ghana and Sub-Saharan Africa. He is a member of the Advisory Board for Journal of *Enterprise Development and Microfinance*. He has been reviewing articles for a number of high-ranking journals.

Sebu Joshua is a Lecturer at the Department of Data Science and Economic Policy at the School of Economics, University of Cape Coast, Ghana. He was awarded his Ph.D. from the University of Kent, UK, in July 2017. He received his B.A. in Economics and Mathematics and M.Phil. in Economics from the University of Cape Coast. He teaches at the undergraduate and graduate levels at the University of Cape Coast. His research interests include financial inclusion, food security and child nutrition, women empowerment, farm household credit access and productivity, poverty and inequality.

Takyi Paul Owusu is a Lecturer at Department of Economics, Kwame Nkrumah University of Science and Technology, Kumasi, Ghana. He obtained his Ph.D. and Master of Arts in Economics from the National Graduate Institute for Policy Studies (GRIPS), Tokyo, Japan in 2020 and 2017, respectively. In addition, he holds Master of Philosophy and Bachelor of Arts in Economics from the University of Cape Coast, Cape Coast, Ghana. His current teaching and research interests are in the areas of Macroeconomics, Monetary economics, Public economics, International trade and finance, Development finance and Economics of Law.

Tawiah-Mensah Jonathan is a Research Associate at the UKRI-GCRF One Ocean Hub, University of Cape Coast, Ghana. He was awarded his M.Phil. from the University of Cape Coast, Ghana, in March 2021. He received his B.A. in Economics and Sociology from the University of Cape Coast. His research interests include digital sociology, political economy, conflict, peace and security, human rights and social movements.

Wiafe Emmanuel Agyapong is holds Master of Philosophy in Economics from the University of Cape Coast. He has over 6-year combined experience in research and teaching. He is a member of the Centre for West African Studies of UESTC. He currently lectures with the Department of

Economics, Ghana Institute of Management and Public Administration. His interest in international economic, financial economic and development. He has published and presented several peer-reviewed academic papers on varied topics on economics.

Yin Elijah Tukwariba is a Lecturer at the Faculty of Law of the University of Cape Coast (UCC) in Ghana, West Africa. Before joining the faculty, he worked as a Social Development Practitioner for 7 years. His intellectual pursuits intersect as a socio-legal researcher, a theorist, and a practitioner. He is an alumnus of the Institute for Global Law and Policy (IGLP) at Harvard Law School (HLS); a recipient of scholarship awards for the Scholars Writing Workshop in Bangkok, Thailand (2018), and the Global Scholars Academy in Geneva, Switzerland (2020). He has attended and presented conferences papers and has number of peer-reviewed articles and chapters in his name.

List of Figures

List of Tables

CHAPTER 1

Introduction

James Atta Peprah, Harold Ngalawa, Thankom Arun, and Evelyn Derera

The finance-growth nexus is one of the most debated topics in the literature. Since the eighteenth century, when classical thinking set the stage for economic theory, several opinions have emerged on the financial system's importance for economic growth. Among some canonical publications, Bagehot (1873) maintained that the financial system in England played a critical role in stimulating industrialisation by facilitating the mobilisation of capital. Schumpeter (1912) argued that well-functioning banks

J. A. Peprah
Department of Applied Economics, College of Humanities and Legal Studies, University of Cape Coast, Cape Coast, Ghana
e-mail: jpeprah@ucc.edu.gh

H. Ngalawa (✉)
College of Law and Management Studies, University of KwaZulu-Natal, Durban, South Africa
e-mail: Ngalawa@ukzn.ac.za

T. Arun
Essex Business School, University of Essex, Colchester, UK
e-mail: t.g.arun@essex.ac.uk

J. A. Peprah et al. (eds.), *Financial Sector Development in Ghana*, Palgrave Macmillan Studies in Banking and Financial Institutions,
https://doi.org/10.1007/978-3-031-09345-6_1

spur technological innovation by identifying and funding entrepreneurs with the best chances of successfully implementing innovative products and production processes. Robinson (1952) observed that financial sector development follows when enterprises grow (enterprise-lead hypothesis). According to this hypothesis, economic development creates demand for certain financial arrangements. Subsequently, the financial system responds to these demands, enterprises grow, and financial sector development is stimulated.

Over the years, from the position of an "over-stressed" determinant of economic growth (Lucas, 1988), the financial sector has become central to economic growth. This has become possible through a new set of the literature (King & Levine, 1993; Levine, 2021; Rajan & Zingales, 1998) which provides unique and deeper insights into the finance-growth nexus. Empirical evidence shows that the financial sector may affect economic growth by improving resource allocation. The findings have supported the financial opening-up process in many countries. Several studies have shown that the accumulation of profits has occurred through financial channels (Krippner, 2005), and the financial sector has gained a "pop-star status", globally (Arun, 2016). Innovation seems essential in maintaining this status, which we have seen in different forms such as options, and derivatives, among others, which involve risk-based decisions.

A large body of recent empirical evidence suggests that financial sector development facilitates economic growth and supports poverty reduction (see, e.g., Durusu-Ciftci et al., 2017; Rewilak, 2017; Zhang & Naceur, 2019). Long-term financing is essential for the demands of Sustainable Development Goals (SDGs) and meets the conditions for a more sustainable global economy that supports the environment and society. Development of the capital market and innovative funding sources provides access to long-term finance and enhances the role of private sector finance and investment in global development. Irrespective of this impressive growth in products and services, various country experiences reveal that as the financial market develops, multiple challenges to growth and development also arise (Caporale et al., 2015; Valickova

---

E. Derera
College of Law and Management Studies, University of KwaZulu-Natal, Pietermaritzburg, South Africa
e-mail: dererae@ukzn.ac.za

et al., 2015). Recently, Zhang and Zhou (2021) show that in the long-run, the welfare-maximising level of financial development is lower than the growth maximising level.

Financial inclusion is seen as a critical feature of financial development (Le et al., 2019). It can be a driver for economic growth, poverty alleviation, and shared prosperity among citizens of a country (Peprah et al., 2021; World Bank, 2013). Among the issues explored in this strand of the literature is how self-discipline based on present bias theory (trade-off between current and future preference), and financial behaviour and attitudes, contribute to access to finance (Bauer et al., 2012; Kostov et al., 2012). The literature reveals that poor financial knowledge should not be taken for granted even in countries with well-developed markets (Lusardi, 2013). Lusardi (2013) further argues that certain groups such as women, the youth, the aged, the unemployed, and those with low education levels often manage with weak levels of financial literacy. A well-designed financial system underscores efforts to foster the inclusion of the majority of the poor and vulnerable groups in society, such as women. Bridging the gender gap in access to finance promotes financial inclusion, leading to sustainable economic development.

Lending methodologies, financing mechanisms, and risk management of banking institutions are central in the financial sector reform. The legal and regulatory framework challenges affect banks' risk management performance. Theoretically, the decision of banks to lend depends on three observed phenomena: first, banks earn substantial profits from off-balance sheet activities and services; second, the critical point in the customer relationship is the loan decision; and third, what is at stake in the loan decision is the expected value of the entire customer relation (Ahtiala, 2005). It is important to stress that banks use a combination of these factors to decide whether to lend to a customer at each point in time. In many countries, small and medium-sized businesses are the backbone of the economy. Across all life cycle stages, these enterprises require access to funding sources for their creation, survival, and growth. How the banking sector reforms address small and medium businesses' structural challenges remains a significant concern.

1.1 Ghana's Financial Sector

Prior to 1988, the Ghanaian government used direct measures of control to intervene in the operations of the financial sector, with the intention of ensuring that certain groups have access to financial services (Aryeetey, 1996). At the time, the formal financial sector in the country was dominated by state-owned banks that enjoyed a monopoly over the entire banking sector in terms of their spread and operations (Biekpe, 2011). There were only two foreign banks in operation, namely Barclays Bank and Standard Chartered Bank (World Bank, 1995). Since 2002, Ghana's bank-based financial system has grown significantly with the influx of many bank and non-bank financial institutions. The rapid growth of Ghana's financial sector has been accompanied by three key developments, namely the introduction of innovative products, revision of regulatory guidelines, and mainstreaming of the microfinance sector. However, IMF (2011) alludes to several vulnerabilities, such as foreign banks' domination, government's involvement in commercial banks operations, lack of access to long-term finance, and weak internal control mechanisms as some of the factors that laid the basis for a structural transformation of the financial sector in Ghana.

Currently, Ghana has 23 registered commercial banks, 144 rural and community banks, 137 deposit-taking microfinance institutions, 25 savings and loans companies, and 31 microcredit institutions in Ghana (Bank of Ghana, 2022). Besides the commercial banks, the financial market in Ghana has insurance companies and some capital markets institutions, which have also recorded remarkable growth. For instance, the number of financial institutions engaged in insurance-related activities has risen to 151 in 2019 (see National Insurance Commission Report 2022a, 2022b). The number of listed companies on the Ghana Stock Exchange has also increased from 7 in 1992 to 13 in 2000, 28 in 2014, 32 in 2018, and 42 in 2022. The GSE was established in July 1989 as a private company limited by guarantee under the Companies Code of 1963.

Three key developments, namely, the introduction of innovative products, revision of regulatory guidelines, and mainstreaming of the microfinance sector, have accompanied the rapid growth of Ghana's financial sector. Currently, Ghana has one of the most active and fastest-growing mobile money markets on the continent, supported by the government's digitalisation agenda (Bala et al., 2022).

From mid-2016, crises hit Ghana's banking and non-banking sectors, which led to the collapse and revocation of institutional licences. Following the crisis, the Bank of Ghana (BoG) embarked on a massive banking sector clean-up, recapitalisation, and other regulatory reforms from mid-2017 to December 2018. These actions align with its mandate to promote the financial safety, soundness, and stability required to foster economic growth and development. The financial disruptions have impacted the dynamism of the overall financial system in the country. The BoG facilitated the takeover of UT Bank Limited and Capital Bank Limited by GCB Limited in August 2017. GCB Bank took over the two banks because of their severe capital impairment and substantial non-performing loans. In 2018, the Central Bank revoked the licences of five commercial banks and merged them into the Bank of Ghana (CBG). In addition, licences for over 343 Microfinance Companies, 28 Micro-Credit Companies, and 23 Savings and Loans Companies were revoked.

A sound risk management system is necessary to support the market participants' confidence in the banking system. Solid legal and regulatory frameworks are essential to a risk management system, and the frameworks require constant revision and harmonisation to capture the emerging challenges in the sector. For instance, the regulatory fabric to support credit risk management issues in Ghana remains scant despite the recent problems in the banking sector, which pose a significant threat to the financial sector's health. Loss of deposits among households and business establishments reduced public confidence in the banking system.

1.2 Focus of the Book

Given that Ghana's financial industry has gone through several metamorphoses in recent times, it is imperative to investigate the relevant policies and practices that would strengthen the gains made in the sector. This book explores the issues of bank stability, financing models, and development challenges in Ghana's financial sector. A holistic approach to ensuring stability in the country's financial sector is required for a sustainable financial market. While adhering to monetary policy effectiveness alongside inclusive finance, developing innovative products and services is critical to ensuring effective financing mechanisms for Small, Medium, and Micro Enterprises (SMMEs). In Chapter 2, Ofori and Obeng (*What Really Drives Financial Sector Development in Ghana?*) use the recent advances in machine learning to present a model for predicting financial development. The authors employed regularisation techniques and identified variables vital for driving and predicting financial development in Ghana. Ofori and Obeng take a position that variables such as economic globalisation, internet access, and institutional quality matter for Ghana's financial sector development.

Ghana has recently embarked on several initiatives to improve the efficiency of its financial sector by resorting to new financial products and services. In Chapter 3, Sebu, Amo-Bondzie, Ewuzie, and Tawia-Mensah (*Harnessing Financial Innovation for Financial Inclusion in Ghana)* describe the nature and extent to which Ghana is harnessing financial innovation to facilitate financial inclusion. The authors present evidence that Ghana has improved its banking operations by using automated teller machines (ATMs), domestic and internationally recognised ATM cards, mobile short codes, and bank-specific mobile application software. However, the most considerable momentum is in the use of mobile money subscriptions and transactions. The chapter also highlights the critical threat of financial innovation fraud and its implications for client protection. The authors call for robust external and internal risk assessment mechanisms to combat such threats to mitigate financial innovation fraud.

Over the last three decades, one key aim of the Ghanaian economy has been to move from a lower-middle-income economy to a higher-income economy in the shortest possible time. In Chapter 4, Cantah, Peprah, and Takyi (*Financial Inclusion and Monetary Policy Effectiveness in Ghana)* examine the role of financial inclusion on the effectiveness of monetary policy in Ghana. The authors employ a structural vector autoregressive (VAR) model by interacting financial inclusion with monetary policy to examine its impact on inflation and output. The chapter estimates the impact of financial inclusion on monetary policy effectiveness.

Women's integration into global finance has been gaining increasing attention in recent efforts towards financial inclusion debates. In Chapter 5, Arun, Annim, Bose, and Arun (*Gendered Financial Behaviour in Ghana: A Comparative Study with South Africa)* examine how gender differences influence different financial services in two sub-Saharan African countries of Ghana and South Africa, and sheds light on the ambiguities of the neoliberal gender agenda in the development discourse. The authors question the social stereotypes on gendered economic behaviour and risk aversion, as it is observed that fewer women in Ghana are likely to use general financial and investment products than women in South Africa. The chapter finds that financial behaviours between men and women in Ghana and South Africa are diverse, and that there is high heterogeneity in using various financial services. The chapter highlights the need for discourse on financial inclusion to include non-financial approaches to understand how gendered social status engenders embodied financial behaviours, social relations, and risk aversion.

Implementation of the Financial Sector Adjustment Programme (FINSAP and the FINSSP I & II) was expected to increase competition and stability in the financial sector. However, since the implementation of these reforms, very little has been done to empirically verify the extent to which a rise in the competitiveness of the banking sector has added to its stability. In Chapter 6, Cantah, Darfor, Nunoo, Afful, and Wiafe (*Bank Competition and Financial Sector Stability in Ghana*) examine the effects of bank competition on financial sector stability in Ghana. The authors argue that the Bank of Ghana (BoG) should continue its work towards strengthening the capital adequacy framework and promote the consolidation of smaller sized banks using mergers and acquisitions. Since the empirical findings provide evidence that a lower level of competition enhances stability in the industry, the consolidation of the banks enhances the market power and supports the banking industry's overall stability.

Several scholars have investigated factors that may explain banking instability in Ghana. One of the distinguishing features among these studies is the measure of banking instability employed. Ngalawa and Derera, in Chapter 7 (*Macroeconomic Determinants of Banking Instability in Ghana*), contribute to the bank instability debate by using a relatively new empirical framework that distinguishes between banking instability caused by bank runs and banking instability emanating from the insolvency problem of banks. The authors argue that an increase in real output reduces the probability of bank runs and banks' insolvency. Higher inflation levels are detrimental to bank stability and can increase the probability of bank runs but decrease banks' insolvency. Contrary to theoretical expectations, the authors found that raising interest rates reduces the likelihood of bank runs and bank insolvency.

The increasing overlap in the operations of formal and informal banks in most developing economies has raised the issue of whether they complement or compete with each other within the financial space. In Chapter 8, Adjei and Ayisi (*Financial Dualism in Ghana: The Implications for Monetary Policy on Loan Variations*) examine the implications of financial dualism for monetary policy in Ghana. The authors confirm the dualistic nature of Ghana's financial sector despite the effort by government and the Central Bank to formalise the informal sector of the economy. The authors find contrasting evidence on the role of monetary policy in determining credit variations of formal and informal banks, raising concerns about the right policy mix and effectiveness of monetary policy control in the informal sector in Ghana.

A pathway to achieving growth in any economy is through access to finance, whether for businesses or consumers. In Chapter 9, Korankye-Sakyi, Abe, and Yin (*Revisiting MSMEs Financing Through Banking Reform Processes: Assessing the Ghanaian Legal Experiences*) examine the institutional and regulatory framework that underpins Ghana's banking reforms and its impacts on access to credit by Micro, Small and Medium Enterprises (MSMEs). The authors opine that, unlike Nigeria and South Africa, the banking reforms in Ghana over the years and their impacts on credit flow to the productive sectors of the economy have not been significant. The chapter advocates for regular revision and harmonisation of the regulations given the finding that an ineffective legal framework contributes to banking sector mismanagement,

In Ghana, where agriculture is the mainstay of the economy, formulating and implementing effective credit risk management policies (CRMPs) to encourage agricultural lending as a viable business is crucial. In Chapter 10, Nyebar, Obalade, and Muzindutsi (*Effectiveness of Credit Risks Management Policies used by Ghanaian Commercial Banks in Agricultural Financing)* investigate the effectiveness of implementing CRMPs used by Ghanaian commercial banks to mitigate credit risk exposure in agriculture financing. The authors note that most Ghanaian commercial banks lack specialised units comprising technocrats with agricultural backgrounds.

Banks have found credit assessment very challenging for a long time. In Chapter 11, Gyimah, Akande, and Muzindutsi (*Evaluation of Lending Methodologies Used by Ghanaian Banks to Extend Credit to SMEs)* examine the lending methodologies used by Ghanaian banks to extend credit to Small and Medium Enterprises. The authors identify two domains of the lending methodologies—Collateral and Banking Records (CBR) and Personal and Business Characteristics (PBCs). Gyimah, Akande, and Muzindutsi argue that the provision of collateral and information on business' policy, philosophy, and performance are outstanding requirements for acquiring credit facilities. The chapter proposes the CBR as a dominant lending method for Ghanaian banks in lending to SMEs.

References

Ahtiala, P. (2005). The new theory of commercial banking and bank lending behavior. *Scottish Journal of Political Economy, 52*(5), 769–792.

Arun, T. (2016). Financialisation in an Interwoven World. In B. Avari & G. G. Joseph (Eds.), *The interwoven world - ideas and encounters in history*, Common Ground Publishing.

Aryeetey, E. (1996). *The formal financial sector in Ghana after the reforms*. Overseas Development Institute.

Bagehot, W. (1873). *Lombard street: A description of the money market*. S. Henry King and Co. http://www.econlib.org/library/Bagehot/bagLomCover.html

Bala, A. R., Behsudi, A., & Owen, N. (2022). Meeting the future. *Finance and Development* , 59(1), IMF.

Bank of Ghana. (2022). Bank of Ghana, Website, accessed on 26 January 2022. https://www.bog.gov.gh/

Bauer, M., Chytilová, J., & Morduch, J. (2012). Behavioral foundations of microcredit: Experimental and survey evidence from rural India. *American Economic Review, 102*(2), 1118–1139.

Biekpe, N. (2011). The competitiveness of commercial banks in Ghana. *African Development Review, 23*(1), 75–87.

Caporale, G. M., Rault, C., Sova, A. D., & Sova, R. (2015). Financial development and economic growth: Evidence from 10 new European Union members. *International Journal of Finance & Economics, 20*(1), 48–60.

Claessens, S. (2006). Access to financial services: A re-view of the issues and public policy objectives. *World Bank Research Observer, 21*(2), 207–240.

Cole, S., Sampson, T., & Zia, B. (2011). Prices or knowledge? What drives demand for financial services in emerging markets? *Journal of Finance, 66*(6), 1933–1967.

National Insurance Commission (Ghana) (2022b). 2019 *Annual report of the insurance industry in Ghana*. National Insurance Commission: Accra. Accessed on 22 March 2022. https://nicgh.org/procurement-notices/annual-report/

National Insurance Commission (Ghana) (2022a). 2010 *Annual report of the insurance industry in Ghana*. National Insurance Commission: Accra. Accessed on 22 March 2022. https://nicgh.org/procurement-notices/annual-report/

Durusu-Ciftci, D., Ispir, M. S., & Yetkiner, H. (2017). Financial development and economic growth: Some theory and more evidence. *Journal of Policy Modeling, 39*(2), 290–306.

IMF (2011). *Ghana financial system stability assessment update*. Accessed on 22 March 2022. https://www.elibrary.imf.org/view/journals/002/2011/131/article-A001-en.xml

King, R. G., & Levine, R. (1993). Finance and growth: Schumpeter might be right. *Quarterly Journal of Economics, 108*, 717–738.

Kostov, P., Arun, T., & Annim, S. (2012). Determinants of access to finance: An investigation into the Mzansi intervention. *European Journal of Development Research, 24*(3), 397–412.

Krippner, G. R. (2005). The financialization of the American economy. *Socio-Economic Review, 3*(2), 173–208.

Le, T. H., Chuc, A. T., & Taghizadeh-Hesary, F. (2019). Financial inclusion and its impact on financial efficiency and sustainability: Empirical evidence from Asia. *Borsa Istanbul Review, 19*(4), 310–322.

Levine, R. (2021). *Finance, growth, and inequality*. IMF Staff Working Paper, WP/21/164.

Lucas, R. E. (1988). On the mechanics of economic development. *Journal of Monetary Economics, 22*, 3–42.

Lusardi, A. (2013) *Insights: Financial capability*. GCFL Global Center for Financial Literacy. www.finrafoundation.org

Peprah, J. A., Koomson, I., Sebu, J., & Bukari, C. (2021). Improving productivity among smallholder farmers in Ghana: Does financial inclusion matter? *Agricultural Finance Review, 81*(4), 481–502. https://doi.org/10.1108/AFR-12-2019-0132

Rajan, R. G., & Zingales, L. (1998). Financial dependence and growth. *American Economic Review, 88*, 559–586.

Rewilak, J. (2017). The role of financial development in poverty reduction. *Review of Development Finance, 7*(2), 169–176.

Robinson, J. (1952). *The generalisation of the general theory, in the rate of interest and other essays*. Macmillan.

Schumpeter, J. A. (1912). *Theorie der wirtschaftlichen entwicklung*. Dunker & Humblot [*The theory of economic development*, 1912, translated by Redvers Opie. Harvard University Press, 1934].

Valickova, P., Havranek, T., & Horvath, R. (2015). Financial development and economic growth: A meta-analysis. *Journal of Economic Surveys, 29*(3), 506–526.

World Bank (2013). *Global financial development report 2014: Financial inclusion* (Vol. 2). World Bank Publications.

World Bank (1995). Ghana: Growth, private sector, and poverty reduction. Report No. 14111-GH.

Zhang, B., & Zhou, P. (2021). Financial development and economic growth in a microfounded small open economy model. *The North American Journal of Economics and Finance, 58*, 101544.

Zhang, R., & Naceur, S. B. (2019). Financial development, inequality, and poverty: Some international evidence. *International Review of Economics & Finance, 61*, 1–16.

CHAPTER 2

What Really Drives Financial Sector Development in Ghana?

Isaac Kwesi Ofori and Camara Kwasi Obeng

2.1 Introduction

At the heart of inclusive growth is an efficient financial sector (World Bank, 2019). This intuition stems from the scholarly contributions of Mckinnon (1973) and Shaw (1973), which highlights the role of financial sector development for effective resource allocation and economic development. While a thriving financial sector can be growth-inducing, there is also the evidence that excessive expansion of the sector can be harmful to the economy[1] (Peprah et al., 2019; Arcand et al., 2012). For instance, Peprah et al. (2019) warn in the case of Ghana that the development of

[1] This can create instability in the financial sector as the sector become fragile and highly susceptible to crises.

I. K. Ofori (✉)
Department of Economics, University of Insubria, Varese, Italy
e-mail: ikofori@uninsubria.it; ikofori@outlook.com

C. K. Obeng
Department of Economic Studies, School of Economics, University of Cape Coast, Cape Coast, Ghana
e-mail: cobeng@ucc.edu.gh

J. A. Peprah et al. (eds.), *Financial Sector Development in Ghana*, Palgrave Macmillan Studies in Banking and Financial Institutions,
https://doi.org/10.1007/978-3-031-09345-6_2

the country's financial sector beyond 70% can drag down growth. Based on this knowledge, it is crucial therefore for policymakers to identify variables key for shaping financial development. Indeed, evidence abounds that the growth of a country's financial sector hinges on a number of factors, markedly, the strength of the country's regulatory institutions, the macroeconomy, external developments, as well as the internal and external developments in the financial system (see, Aluko & Ajayi, 2018; Feyen et al., 2013; Law & Demetriades, 2006). Despite this knowledge, policymakers, more often than not, grapple with the problem of identifying which of the aforementioned factors drives their financial sector development.

More concerning is the inconclusive results on the effect of these variables on financial development (see e.g., Almarzoqi et al., 2015; Marcelin & Mathur, 2014). The inconclusive results are due to two shortfalls. First, the use of variables such as the proportion of financial institutions' assets to gross domestic product (GDP) and the share of commercial banks' total deposits to GDP as proxies for financial sector development (see e.g., Mtar & Belazreg, 2020; Barajas et al., 2013; Adu et al., 2013). Second is the preferential selection of covariates for inference or prediction from a pool of potential predictors determinants of financial development in regression problems (see e.g., Aluko & Ajayi, 2018; Ibrahim & Sare, 2018). The first shortfall has been addressed quite comprehensively by Svirydzenka (2016) who developed the Global Financial Development Index.[2] Since then, a number of studies have utilised the index for both country-specific and cross-country studies (see e.g., Nguyen et al., 2020; Peprah et al., 2019).

Despite the immense progress, rigorous empirical works addressing the second is hard to find. Particularly, the question of which factors are peculiarly germane for Ghana's financial sector development remains unanswered because of preferential selection of predictors evident in prior contributions such as Adu et al. (2013) and Takyi and Obeng (2013). Indeed, knowledge of this is crucial as it informs appropriate policy actions, particularly, on the likely (in)effectiveness of monetary policy transmission mechanism and/or resource allocation. One way of addressing the aforementioned gaps is through the application of machine learning techniques (see, Tibshirani, 1996; Hastie et al., 2001; Zou,

[2] The Global Financial Development index is multidimensional index providing estimates on the level of developments of a country's financial institutions and markets.

2006; James et al., 2013; Chernozhukov et al., 2015). As Tibshirani (1996), Hastie et al. (2001) and Zou (2006) indicate, machine learning regularisation algorithms can be trained to study patterns in large datasets and reduce modelling complexity while yielding spare results.

This forms the motivation of this chapter where machine learning techniques are employed to achieve two main objectives—first, for identifying variables key for driving Ghana's financial sector development, and second, for identifying a model best for predicting financial development in Ghana. The contribution we offer is particularly invaluable to the Government of Ghana's bid of raising the current financial inclusion level of 58% to 85% by 2023 (Government of Ghana, 2018). More germane, our contribution can be an eye-opener for the Government of Ghana and other stakeholders in the financial sector in terms of areas to prioritise in their bid to build a robust and dynamic financial system.[3]

The rest of the Chapter is organised in what follows. Section 2.2 focuses on the overview of Ghana's financial sector while Section 2.3 presents the data and methods underpinning our contribution. Section 2.4 presents the results while Section 2.5 concludes with some policy implications.

2.2 Literature Survey

2.2.1 Overview of Financial Development in Ghana

Ghana's financial sector comprises universal banks, and non-bank financial institutions, chief among them being rural and community banks, microfinance institutions, fund management institutions, pension funds, insurance companies, leasing companies, remittances companies, savings and loans and the Ghana stock exchange. Together with the national insurance commission for insurance, the national pension regulatory authority for pensions, and the securities and exchange commission for the capital market, the Bank of Ghana (BoG) spearheads the licensure, supervision and regulation of the financial system.

Before 1986, Ghana's financial sector was state-monopolised, repressed and shallow (i.e., bank-dominated). The financial sector suffered severely

[3] In line with this agenda is the enumeration of five broad themes targeting: financial stability; access, quality and usage of financial services; financial infrastructure; financial consumer protection; and financial capacity (see, Government of Ghana, 2018).

from the post-independence-controlled regime of the country. The controlled regime was in line with Ghana's industrialisation policy, culminating in interest controls and credit ceilings to ensure that cheap credit was available to government-imposed priority sectors such as the manufacturing sector. The financial sector was also burdened with high taxes, non-performing loans and high reserve requirements (Aryeetey et al., 2000). The restrictive policies created major distortions and in addition to macroeconomic instability, real interest rates became negative. The negative repressive policies were evident as financial depth plummeted from 24% in 1977 to 12% in 1984 (Aryeetey et al., 2000). The GoG, with the support of the World Bank, introduced the financial sector adjustment programme (FINSAP) in 1987 to revive the distressed financial sector to enhance competition, consumer confidence, affordable financial services and products and effective monetary transmission.

The reforms resulted in the liberalisation of interest rates and the deregulation of the banking industry in general. The reforms also gave birth to the development and expansion of the financial sector through the establishment of the Ghana Stock Exchange, foreign exchange bureaux and non-bank financial institutions with the broad objective of spurring competition in the sector through alternative channels of intermediation. Despite all these achievements, the financial sector was saddled with macroeconomic instability and low investment prior to the early 2000s. Despite these challenges, Ghana's financial sector has grown remarkably since 2010, but it still remains bank-dominated. Total financial sector assets grew from 53% of GDP in 2010 to 78% in 2017. In the last decade, Ghana's financial sector has also witnessed a proliferation of unlicensed/illegal deposit-taking financial institutions[4] and fraudulent online trading activities[5] that resulted in either a collapse, merger or downgrade of some financial intermediaries. This necessitated a clean-up in 2017 and 2018 aimed at restoring consumer confidence in the sector. The clean-up resulted in 23 universal banks, down from 32. While the privately owned institutions lost their numerical strength, the state-owned increased from 3 to 4. There are currently 19 privately owned banks, with 17 of them foreign-owned. To put the overview of Ghana's financial sector into perspective, Fig. 2.1 is presented. Indeed, Fig. 2.1 suggests

[4] Examples are DKM, God is love; MenzGold and Golden Life Microfinance Limited.

[5] Examples are Hour Forever, BitDax and CriptonBit.

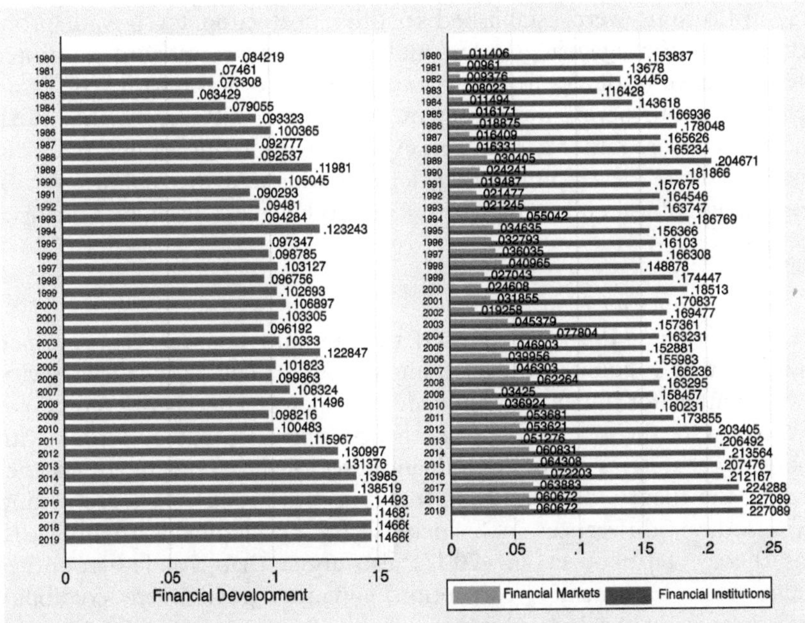

Fig. 2.1 Average financial development (left), and financial markets and institutions (right), 1980–2019, IMF financial development index data

that Ghana's financial market is significantly less developed compared to its financial institutions despite significant strides since 2010.

2.2.2 *Theoretical and Empirical Literature Review*

In this section, we present some theories and empirical works to justify the inclusion of variables identified as potential drivers of financial development in this chapter.

2.2.2.1 *Endowment Theory (Settler Mortality Hypothesis)*

The endowment theory as put forward by Acemoglu et al. (2001) underscores the relevance of institutions, resource endowment and geography for financial sector development. The theory indicates that in the colonial era, institutions were established to offer protection for private property; protection against government power of expropriation; guarantee the transfer of resources from the colony to the coloniser with little or no investment. Furthering the argument, Beck et al. (2003) argue that initial endowments are rather germane in explaining international differences in financial sector development than legal origins and that countries with poor geographical endowments are likely to have less developed financial sectors.

2.2.2.2 *Law and Finance Theory*

La Porta et al. (1998) introduced the law and finance theory, which suggests that a country's legal framework (particularly, financial sector laws) matter for financial sector development. The theory comes in twofold. On the one hand, the theory suggests that countries with robust legal systems realise high financial sector development. In another breath, the theory identifies legal traditions[6] as the driving force behind cross-country differences in financial sector development. In line with this theory, Djankov et al. (2007) also argue that civil law countries realise lesser bureaucracy, corruption, enhanced government credibility and greater financial development. A more recent work in Fowowe (2014), however, does not find support for the theory in the case of some African countries.

2.2.2.3 *Financial Liberalisation Theory*

The theory follows the arguments by McKinnon (1973) and Shaw (1973) that a country's financial sector develops following a liberalisation from government repression. Additionally, the theory indicates that both domestic savings and the advancement of loanable funds for investment increase if there is a moderately high and positive interest rate. They argue that financial repression results in market disequilibrium, consequently limiting allocative efficiency due to credit controls and limited access to external finance. Furthering the thrust of this theory, Baltagi et al. (2009) argue that a country's financial sector grows even faster if financial liberalisation is accompanied by greater trade and financial openness.

[6] La Porta et al. (1998) argue that, common law countries provide stronger legal protection for investors than civil law countries.

2.2.2.4 *Inflation and Finance Theory*

This theory was put forward by Huybens and Smith (1999). The theory posits an inverse relationship between inflation and financial intermediation. The authors argue that macroeconomic instability, evidenced by rising inflation causes financial institutions to ration credit, reducing financial market activity and profitability. Furthering the import of this theory, Rousseau and Wachtel (2002) provide evidence to show that high inflation discourages long-term loans, resulting in inefficient allocation of resources.

2.2.2.5 *Growth-Led Hypothesis*

The hypothesis follows the argument by Robinson (1952) that rising economic expansion is a significant driver of demand for financial services. The theory suggests that rising living standards is crucial for achieving greater financial inclusion and development. The theory has been enhanced tremendously with studies such as Liang and Jian-Zhou (2006), providing empirical evidence. It follows that, in developing or emerging economies, achieving sustained economic growth can foster financial sector activity, profitability and development.

2.2.3 *Determinants of Financial Sector Development*

Several empirical works have identified variables crucial for driving financial development. In this section, some of these empirical works are presented.

Empirical evidence shows a strong positive relationship between financial sector development and savings mobilisation. The argument stem from the notion that when countries liberalise interest rates, it enhances financial deepening (McKinnon, 1993; Shaw, 1973). Similarly, Mavrotas and Santillana (1999) present the theoretical links between financial sector liberalisation and savings mobilisation based on the life cycle or permanent income theory of consumption. It is argued that liberalising the financial sector can spur competition among providers of financial services, eliminating all forms of borrowing constraints. In effect, it allows individuals to borrow to smoothen their consumption patterns.

Additionally, prior empirical works such as Ayadi et al. (2015), Elsherif (2015), Sanusi et al. (2017) and Aluko and Ibrahim (2019) identify factors such as inflation and public debt as financial development impediments. For instance, Ayadi et al. (2015) provide evidence to argue that soaring public debt does not only impedes the growth of credit but also crowds out private investment. Similarly, Boyd et al. (2001) also argue that inflation can be detrimental to the robustness and efficiency of financial markets and institutions. This is particularly pronounced in inflation-targeting countries where rising inflation levels create frictions in the credit market, triggering financial sector setbacks in the process (Padachi et al., 2008). Empirical evidence to this effect is seen in Bittencourt (2011) who find that inflation impedes financial development in Brazil.

Evidence also shows that sound economic, legal and institutional governance also matters for financial development. For instance, Khalfaoui (2015) and Shabbir et al. (2018) reckon that fiscal discipline, sustained economic growth and regulatory quality are essential for boosting financial development. Similar work is found in Beck and Levine (2005) who argue that prudential supervision enhances financial liberalisation policies and eventual development of financial systems.

On the implication of political governance for financial sector development, Naqvi et al. (2017) provide evidence to show that sociopolitical instability hampers financial sector development. In a similar vein, empirical contributions by Cherif and Dreger (2016) also suggest that financial reforms, rule of law or democratic governance are crucial for promoting financial development. This evidence is closely linked with that of Voghouei et al. (2011) and Khalfaoui (2015), which indicates that political and legal traditions are key determinants of financial development in the developing world.

In exploring the determinants of financial development in Egypt, Elsherif (2015) applied the ARDL and Johansen test for Cointegration to a time series data spanning 1974–2012. The author provided strong evidence to show that variables such as economic growth, trade openness, investment, human capital and per capita income drive financial development. A similar empirical contribution is seen in Boopen et al. (2011) who examined the determinants of financial development in Mauritius using the ARDL approach. The results revealed that covariates such as trade liberalisation, investment rate, per capita income and educational attainment are important determinants of financial development. In a more comprehensive work, Aluko and Ibrahim (2019) employed the two-step system GMM to a panel of 32 countries in sub-Saharan Africa and found that trade openness and the level of income positively affect the development of financial institutions. In the same fashion, Demetriades and Hook

Law (2006) employed dynamic panel data analysis for 43 developing countries for the period 1980–2001 and found that institutional quality, as well as trade and capital account openness, is key for a well-developed financial sector.

On the contrary, David et al. (2014) did not find evidence of the economic globalisation-financial development relationship in the case of sub-Saharan Africa. We sight similar evidence in Bandura (2020) and Aluko and Ibrahim (2020) who point out that trade openness has no direct impact on financial development. However, both authors made an interesting observation that it is only when a country has better institutions that trade openness have a significant impact on financial development. Similarly, Asongu (2012) provides strong evidence to show that greater openness to international commodity and financial markets impede bank efficiency in 29 low- and middle-income African countries.

Filippidis and Katrakilidis (2014) examined the effect of fiscal policy on financial development in 39 low- and lower-middle-income countries and provided robust evidence to support the claim that government expenditure boosts the development of financial institutions. In a similar empirical work, Aluko and Ibrahim (2019) find that government expenditure enhances the development of financial institutions in 32 sub-Saharan African countries. However, the political economy view points to the suppressing effect of government expenditure on the competition, efficiency and development of financial institutions. Evidence to this view is the work of Cooray (2011) which provides evidence from a sample of 71 countries to suggest that uncontrolled government expenditure is detrimental to the efficiency of financial institutions. Further, Naceur et al. (2014) and Mahawiya (2015) have tested the classical dictum that excessive government spending crowds out private investment and the progress of the financial sector. The results from these studies indicate that high government consumption expenditure is deleterious to the growth and sustainability of financial institutions' development in the Mediterranean and sub Saharan Africa. A summary of studies exploring the drivers of financial development is provided in Table 2.1.

Table 2.1 Summary of studies on drivers of financial developments

Authors	*Topic*	*Study area*	*Approach*	*Results*	*Drivers of financial development*
Naceur et al. (2014)	What drives the development of the MENA financial sector?	MENA countries	Fixed-effect and random-effect models	Government expenditure affects financial sector development either through competition or infrastructural development	Government expenditure, infrastructure
Aggarwal et al. (2011)	Do remittances promote financial development?	Cross country	System GMM	Remittances improve bank deposits, credit and financial sector development	Remittances
Peprah et al. (2019)	Financial development, remittances and economic growth: a threshold analysis	Ghana	ARDL	Remittances increase the volume of deposits, credit and financial sector development	Remittances
Boyd et al. (2001)	The impact of inflation on financial sector performance	Cross country	OLS	Inflation hinders both banking sector development and equity market development	Inflation
Bittencourt (2011)	Inflation and financial development: evidence from Brazil	Brazil	OLS	Inflation has detrimental effects on financial development	Inflation
Ahmed (2013)	Effects of financial liberalisation on financial market development and economic performance of the SSA region: an empirical assessment	sub-Saharan African	System GMM	Financial globalisation hampers financial development	Financial liberalisation

Authors	*Topic*	*Study area*	*Approach*	*Results*	*Drivers of financial development*
Boopen et al. (2011)	Determinants of financial development: the case of Mauritius	Mauritius	ARDL	Trade openness, financial liberalisation, financial development, investment rate, per capita and literacy rates are factors stimulating financial development while inflation adversely influence financial development both in the short and long run	Trade openness, financial liberalisation, investment rate, per capita and literacy rates and inflation
Demetriades and Hook Law (2006)	Openness, institutions and financial development	Cross country	Dynamic panel	Institutional quality drives financial development	Institutional quality
David et al. (2014)	Does openness matter for financial development in Africa?	SSA	ARDL	Both trade and capital account openness induce financial development	Trade openness, institutional quality, GDP per capita and inflation
Bandura (2020)	Trade openness, institutions and financial development in sub-Saharan Africa	SSA	GMM	Trade openness has no significant effect on financial development but there is a positive effect of institutional quality on the development of the financial sector	institutional quality

(continued)

Table 2.1 (continued)

Authors	*Topic*	*Study area*	*Approach*	*Results*	*Drivers of financial development*
Asongu (2012)	How has mobile banking stimulated financial development in Africa?	African countries	Two-stage least squares	Mobile banking has a negative incidence on traditional financial intermediary dynamics of depth, activity and size	Mobile banking
Cooray (2011)	The role of the government in financial sector development	Cross country	IV regression	Governance quality enhances both financial sector size and efficiency. Government expenditure and State banks impede financial sector efficiency	Quality of government and government expenditure
Khalfaoui (2015)	The determinants of financial development: empirical evidence from developed and developing countries	Cross country	Fixed-effect and random-effect models	Economic stability and the legal and institutional framework have a significant impact on financial development	Economic stability, legal and institutional
Naqvi et al. (2017)	Impact of political instability on financial development of Pakistan	Pakistan	OLS	Political instability has negative significant impact on financial development while Legal protection and GDP per capita positively influence financial development	Political instability, trade openness, Legal protection and GDP per capita

Authors	*Topic*	*Study area*	*Approach*	*Results*	*Drivers of financial development*
Voghouei et al. (2011)	A survey of the determinants of financial development	Cross country	Systematic review	Institutions, openness of trade and financial markets, legal tradition and political economy promote the financial development	Institutions, openness of trade and financial markets, legal tradition and political economy
Kodila-Tedika and Asongu (2015)	The effect of intelligence on financial development	Cross country	OLS	Positive correlation between Human capital measured in terms of IQ, cognitive ability & cognitive skills and financial development	Human capital measured in terms of IQ, cognitive ability and cognitive skills
Opperman and Adjasi (2019)	Remittance volatility and financial sector development in sub-Saharan African countries	SSA	Two-step system GMM estimator	Remittance volatility, corruption, rule of law income per capita, and trade openness and inflation, drive financial development	Remittance volatility, corruption, rule of law income per capita and trade openness and inflation
Tayssir and Feryel (2018)	Nexus between central bank and financial development	Cross country	System GMM	Central bank influence level of financial development	Central bank
Baltagi et al. (2009)	Financial development, openness and institutions: evidence from panel data	Cross country	Dynamic panel estimation techniques	Openness and institutions influence financial sector development	Openness and institutions

(continued)

Table 2.1 (continued)

Authors	*Topic*	*Study area*	*Approach*	*Results*	*Drivers of financial development*
Coulibaly (2015)	Remittances and financial development in sub-Saharan African countries: a systems approach	SSA	Panel Granger causality approach	Remittances do not promote financial development	Remittances
Aluko and Ibrahim (2019)	On the macroeconomic determinants of financial institutions development in sub-Saharan Africa	SSA	System GMM	Trade openness, income and government expenditure are positive determinants of financial institutions development while higher inflation inhibits improvement in financial institutions	Trade openness, income, and government expenditure, and inflation
Ayadi et al. (2015)	Determinants of financial development across the Mediterranean	Northern and Northern Mediterranean countries	Random-effect	Strong legal institutions, good democratic governance and adequate implementation of financial reforms and capital inflows positively impact on financial development. While inflation, government debt growth negatively affects financial development	Strong legal institutions, good democratic governance, adequate implementation of financial reforms, capital inflows, inflation and government debt growth

Source Authors' construct (2021)

2.3 Data and Methodology

2.3.1 Data

The chapter employs time series macro data covering the period 1980–2019 for the analysis. In this chapter, a total of 81 potential determinants of financial development are drawn for the analysis. These covariates comprise several bank-specific indicators, institutional and policy effectiveness scores, socioeconomic indicators and external factors such as exchange rate, interest rate, macroeconomic management and economic growth. Data on the variable of interest (i.e., financial development), bank-specific and external indicators of the financial system are sourced from the International Monetary Fund's Global Financial Development Index (Svirydzenka, 2016), and the World Bank's Global Financial Development Database (Čihák et al., 2013). While data on poverty and income distribution are sourced from the Global Consumption and Income Project (Lahoti et al., 2016), we draw that of globalisation from the KOF globalisation index (Gygli et al., 2019). Furthermore, to capture the effect of structural, institutional, and macroeconomic variables such as exchange rate, poverty, unemployment, economic growth, inequality and telephone subscription are taken from the World Development Indicators. In Table 2.2, a description of all variables used in the chapter is presented.

2.3.2 Estimation Strategy

In this section, the empirical foundation of the chapter is presented. The first part of the section (i.e., Sect. 2.3.3) looks at the specification of the shrinkage (variable selection) techniques while that of the inferential models are presented in Sect. 2.3.4. In recent times, the debate on whether classical liner estimators such as the ordinary least squares or machine learning regularisation techniques are appropriate for inference or prediction is gaining attention. Researchers in favour of the former argue that the right covariates for empirical analysis can be chosen based on relevant theories. However, researchers employing machine learning algorithms in works of this nature argue that classical techniques may not yield sound estimates in regression problems where the number of covariates exceeds observations. Even in the extreme cases where this is possible, the presence of too many covariates like those underpinning this Chapter (i.e., 81) may cause overfitting. This arises because as the features become large, least-squares assumptions of no multicollinearity, homoscedasticity

Table 2.2 Variable definition and data sources

Variables	*Definition*	*Source*
Grosssav	Adjusted annual gross savings (% of GNI)	WB
Econ_mgt	Economic management policy effectiveness rating (1 = low to 6 = high)	WBFD
Crev_efficiency	(1 = low to 6 = high)	WBFD
Cfinsector_rating	Financial sector management rating (1 = low to 6 = high)	CPIA
Cmacro_mgt	Macroeconomic management rating (1 = low to 6 = high)	CPIA
Cpolicy_Institu	Policy and institutions for environmental sustainability rating (1 = low to 6 = high)	CPIA
Cpublic_sector mgt	Public sector management and institutions cluster average (1 = low to 6 = high)	CPIA
Cps	Private credit by deposit money banks and other financial institutions to GDP (%)	WB
Ease_score	Ease of doing business index (1 = most business-friendly regulations)	WB
Educ_uppersec	Number of people with at least High School education	WB
Service_emp	Service sector employment (people)	WB
Agric_emp	Agricultural sector employment (people)	WB
Ind_emp	Industrial sector employment (people)	WB
Bbs	Fixed broadband subscription per 100 people	WBFD
Tel_ssc	Telephone subscription per hundred people	WBFD
Fte_ sus	Fixed telephone subscription per hundred people	WBFD
Fdi	Net inflow of FDI (%GDP)	WB
Gdpg	Annual GDP growth rate	WB
Gini	Income inequality indicator (Gini index)	WB
Gfcfg	Gross fixed capital formation growth	WB
Hci	Human Capital Index (HCI) (scale 0–1)	WB
Money_supply	Money supply growth	WBFD
Internet	Secure internet servers (per 1 million people)	WBFD
Inflation	Consumer price index (2010 = 100, average)	WB
Insufin_serv	Institutions offering insurance and financial services	WBFD
Celfon_sus	Cellphone subscription	WBFD
Povglw	Poverty gap at \$1.90 a day (2011 PPP) (%)	WB
Povghi	Poverty gap at \$3.20 a day (2011 PPP) (%)	WB
Profittax	Tax on profits of financial institution's profit	WBFD
Rer	Real effective exchange rate	WB

(continued)

Table 2.2 (continued)

Variables	*Definition*	*Source*
Self_emp	Workers engaged in own-account work, producers' cooperatives, and contributing family workers	WB
Taxgdp	Tax revenue (%GDP)	WB
Trade	Sum of import and export (%GDP)	WB
Unemp	Unemployment, total (% of total labour force)	WB
Z-score	Bank Z-score (financial institutions stability)	WBFD
Boone	Boone indicator (financial sector competition)	WBFD
Firmsbank_invest	Firms using banks to finance investments (%)	WBFD
Firmbank_wcap	Firms using banks to finance working capital (%)	WBFD
Firmsacc	Firms with a checking or savings account (%)	WBFD
Foreignbanks	Number of foreign banks in the	WBFD
Remit	Inflow of remittance as a share of GDP	WBFD
Stockmktcap	Stock market capitalisation to GDP (%)	WBFD
Stockpvol	Stock price volatility index	WBFD
FD	Financial development index	FD Index
Bankasset	Bank liquid reserves to bank asset ratio (%)	WBFD
Acc fininsti	Bank account in any financial institutions	WBFD
Acc remit	Bank account for receiving remittances	WBFD
Acc wages	Bank account for receiving wages and salaries	WBFD
ATM	Availability of automated teller machines	WBFD
Bank_acc	Bank accounts per 1,000 adults	WBFD
Bank_branches	Bank branches per 100,000 adults	WBFD
Bank_capasset	Bank capital to total assets (%)	WBFD
Bank_concentrat	Bank concentration (%)	WBFD
Bank_costtoincratio	Bank cost to income ratio (%)	WBFD
Bank_credittodep	Bank credit to bank deposits (%)	WBFD
Bank_depgdp	Bank deposits to GDP (%)	WBFD
Irs	Bank lending-deposit spread	WBFD
Nim	Bank net interest margin (%)	WBFD
Bank_noninterestinc	Bank noninterest income to total income (%)	WBFD
Npl	Bank non-performing loans to gross loans (%)	WBFD
Bank_overheadcosts	Bank overhead costs to total assets (%)	WBFD
Bank_regucapriskasset	Bank regulatory capital to risk-weighted assets (%)	WBFD
Roa_net	Bank return on assets (%, after tax)	WBFD
Roa_gross	Bank return on assets (%, before tax)	WBFD
Roe_net	Bank return on equity (%, after tax)	WBFD

(continued)

Table 2.2 (continued)

Variables	*Definition*	*Source*
Roe_gross	Bank return on equity (%, before tax)	WBFD
Urban_pop	Annual population growth rate (urban)	WB
Population_growth	Annual population growth rate (overall)	WB
House_spend	Growth rate in household consumption expenditure	WB
Gov_gdp	Government expenditure as a percentage of GDP	WB
Gpc	GDP per capita (2017 PPP)	WB
Telefon	Fixed telephone subscription (1 million of the population)	WBFD
Kofgi	KOF. overall globalisation index	KOF. Index
Kofec	KOF. economic globalisation index	KOF. Index
Koftr	KOF. trade globalisation index	KOF. Index
Koffin	KOF. financial globalisation index	KOF. Index
Kofso	KOF. social globalisation index	KOF. Index
Palma	Income inequality as measured by the Palma ratio	GCIP
Theil	Income inequality as measured by the Theil index	GCIP
Gdpgsqr	GDP growth squared	Generated

Note FD Index is Financial Development (International Monetary Fund); WBFD is the World Bank Global Financial Development Database; KOF. Index is the Konjunkturforschungsstelle (KOF) index; GCIP is Global Consumption and Income Project;
CPIA is Country Policy and Institutional Assessment (World Bank); and WB is World Bank's World Development Indicators
Source Authors' construct (2021)

and exogeneity typically break down. Though overfitting does not bias the resultant, it renders inferences and predictions flawed (James et al., 2013).

It is in the remit of these concerns that the popularity of machine learning techniques is rising. More importantly, machine learning techniques can be applied irrespective of the number of predictors, nonlinearity, model specification and time (Tibshirani, 1996). In this chapter, therefore, we employ recent advances in regularisation to learn patterns in the underlying dataset to yield sparse predictors of financial development though at a cost (i.e., the bias associated with variable selection). That said, we train three shrinkage models from the lasso family—the Standard lasso, the Minimum BIC lasso and the Adaptive lasso to achieve the first objective.[7] We proceed to perform causal inference on the selected variables in Objective 1 by employing the partialing-out lasso linear regression, double-selection linear lasso, and partialing-out lasso instrumental variable regression. Finally, we point out that STATA (v16) and R (v3.6) software are relied upon for the analysis. While the latter is used for data imputation and generating summary statistics, the former is employed for data partitioning, regularisation and inference.

2.3.3 *Specification of Variable Selection Models*

2.3.3.1 *Specification of Standard Lasso and Minimum BIC Lasso Models*

Tibshirani (1996) introduced the Standard lasso to address the ineffectiveness of classical estimation techniques for subset selection in large datasets. The main advantages of the Standard lasso like other shrinkage methods are that it: (1) enhances the model interpretability by weeding out weak predictors of the outcome variable; (2) enhances the effectiveness of prediction as regularisation reduces variance without a substantial increase in the bias; and (3) is limitless to data dimensionality. In this chapter, therefore, the *Standard lasso* is employed to select the key drivers of financial development by penalising the model coefficients through a tuning parameter (λ). In line with Tibshirani (1996), we specify the objective function of the Standard lasso as seen in Eq. (2.1). For the Standard lasso algorithms to 'catch' the key covariates of financial development from the 81 potential predictors, the penalty ($\lambda \sum_{j=1}^{\rho} |\beta_j|$), also referred to ℓ_1-norm, is introduced to obtain $\widehat{\beta}_{lasso}$ as shown in Eq. (2.2)

$$Q_L = \frac{1}{N} \sum_{i=1}^{N} \omega_i f(y_{it}, \beta_0 + X_{it}\beta') + \lambda \sum_{j=1}^{p} k_j |\beta_j| \tag{2.1}$$

$$\hat{\beta}_{lasso} = min \left\{ SSE + \lambda \sum_{j=1}^{\rho} |\beta_j| \right\}, \tag{2.2}$$

[7] Since the ordinary least squares technique and Ridge regression cannot yield variable selection, their estimations are relaxed.

where y_{it} is financial development, $\boldsymbol{i}$ is country, $\boldsymbol{t}$ is year, X_{it} denotes a vector of 81 possible determinants of financial development. Thus, variable selection occurs by minimising the model sum of square errors with a given ℓ_1-norm. It is imperative to point out that if the tuning parameter, $\lambda = 0$, then no variable selection occurs, while $\lambda \rightarrow \infty$ indicates an intercept-only model (i.e., complete shrinkage). For brevity, we point out that the specifications of the *Minimum BIC* lasso are as specified in Eqs. (2.1) and (2.2) but variable selection decision is based on the Schwarz Bayesian information criterion (Schwarz, 1978).

2.3.3.2 *Specification of Adaptive Lasso Model*

Zou (2006) introduced the Adaptive lasso method to enhance the consistency of regularisation as features grow rapidly. This is done by invoking the '*oracle property*' (z_j) in addition to the ℓ_1-norm. That said, we specify the objective function and the attendant estimator of the Adaptive lasso (i.e., $\hat{\beta}_{AdaptiveLasso}$) as apparent in Eqs. (2.3) and (2.4), respectively.

$$Q_L = \frac{1}{N}\sum_{i=1}^{N} \omega_i f(y_{it}, \beta_0 + X_{it}\beta') + \lambda \sum_{j=1}^{p} k_j |\beta_j| \tag{2.3}$$

$$\hat{\beta}_{AdaptiveLasso} = min\left\{ SSE + \lambda \sum_{j=1}^{\rho} z_j |\beta_j| \right\} \tag{2.4}$$

where y_{it}, $\boldsymbol{i}$, $\boldsymbol{t}$ and X_{it} are as defined above.

2.3.3.3 *Choice of Tuning Parameter*

A fundamental decision regarding variable selection is the choice of the tuning parameter ($\boldsymbol{\lambda}$). The relevance of $\boldsymbol{\lambda}$ is that it controls the shrinkage process. As Schneider and Wagner (2012) reckon, a good tuning parameter is essential for the overall performance regularisation models. Among the widely used methods for choosing an efficient $\boldsymbol{\lambda}$, the Akaike Information Criterion (AIC), Bayesian Information Criterion (BIC) and Cross-validation (CV) are the most widely used (see Tibshirani & Taylor, 2012). A caution regarding $\boldsymbol{\lambda}$ is that if it becomes sufficiently large, it can shrink covariates excessively, pushing relevant variables to zero. Particularly, Zou et al. (2007) indicate that relative to the CV, the BIC and AIC are less volatile and faster compute in small samples. However, to the extent that setting $\boldsymbol{\lambda}$ under a researcher's discretion can yield *target sparsity* and harm both predictive capacity and inferencing (Hastie et al., 2009), we invoke the 10-fold cross-validation to choose the appropriate $\boldsymbol{\lambda}$.

2.3.4 Specification of Lasso Inferential Models

This section presents models used for providing inferences on the selected variables in Objective 1. To this end, we employ the partialing-out lasso linear regression (POLR), the double-selection lasso linear regression (DSL) and the partialing-out lasso instrumental-variables regression (POIVLR), where the variables identified in Objective 1 are used as the variables of interest, and all the unselected drivers become the controls. It is imperative to note that due to the sparsity of the regularisation techniques, the control variables are usually many. Additionally, it is worth noting that aside from the variables of interest, the DSL, POLR and POIVLR do not provide estimates on all other variables. Finally, the appropriateness of the aforementioned techniques rests in the fact that they are consistent and efficient regardless of data dimensionality, time horizon, endogeneity, heteroscedasticity or multicollinearity.

2.3.4.1 Specification of DSL Model

In the remit of Objective 2, we first follow Belloni et al. (2014) the DSL model as apparent in Eq. (2.5):

$$E[y|d,x] = \psi\alpha' + \phi\beta', \tag{2.5}$$

where financial development is denoted by $\boldsymbol{y}$; $\boldsymbol{\psi}$ is composed of $\boldsymbol{J}$ lasso selected key determinants of financial development, and $\boldsymbol{\phi}$ is a vector of $\boldsymbol{p}$ controls strictly taken from the unselected determinants of financial development. It is worth noting that while estimates on $\boldsymbol{J}$ are produced by the DSL, those for $\boldsymbol{p}$ are relaxed.

2.3.4.2 Specification of POLR Model

An added advantage of the POLR over the DSL is that it improves the consistency of inference as the model becomes too complex. To this end, we follow the POLR specification of Belloni et al. (2012) and Chernozhukov et al. (2015) as apparent in Eq. (2.6).

$$E[Y|d,x] = d\alpha' + X\beta', \tag{2.6}$$

where $\boldsymbol{y}$ is financial development, $\boldsymbol{d}$ is a vector of the $\boldsymbol{J}$ financial development determinants of interest, and $\boldsymbol{X}$ contains the $\boldsymbol{p}$ controls of financial development. As we pointed out in the case of the DSL, the POLR reports coefficients, standard errors and confidence intervals on the $\boldsymbol{J}$ variables alone.

2.3.4.3 Specification of POIVLR Model

Unlike the DSL and POLR, the POIVLR can yield unbiased and efficient estimates in the presence of endogeneity. In a high dimensional dataset like the one employed in this chapter, sources of endogeneity abound. For example, endogeneity can arise due to the bi-causal relationship between economic growth and financial development as argued succinctly by McKinnon (1973) and Shaw (1973). To ensure that our empirical estimates are unbiased and inefficient, we follow Chernozhukov et al. (2015) by applying the POIVLR technique. We specify the POIVLR as:

$$y = \Psi\alpha_d' + \Phi\alpha_f' + X\beta' + \varepsilon \tag{2.7}$$

where $\boldsymbol{y}$ is financial development; $\boldsymbol{\Psi}$ encompasses $\boldsymbol{J_d}$ endogenous variables of interest; $\boldsymbol{f}$ comprises the $\boldsymbol{J_f}$ exogenous variables of interest; and $\boldsymbol{X}$ contains $\mathscr{P}_x$ controls, $\mathscr{P}_z$ outside instrumental variables denoted by z that are correlated with $\boldsymbol{d}$ but not with $\boldsymbol{\varepsilon}$ are introduced. As aforesaid, the simultaneity between financial development and economic growth presents endogeneity concerns which are addressed using the z instruments.[8]

2.3.5 Data Engineering and Partitioning

Against the backdrop of some missing observations in the underlying dataset, we employ the K-nearest neighbour (KNN) data imputation technique to obtain a fully balanced dataset. In other words, we employ the KNN technique such that we do not run the analysis based on 96.1% observations (see Fig. 2.2) but 100% observations (see Fig. 2.3). That said, we perform a 50–50 split of the dataset based on the Stratified

[8] 48 covariates are specified as potential instruments.

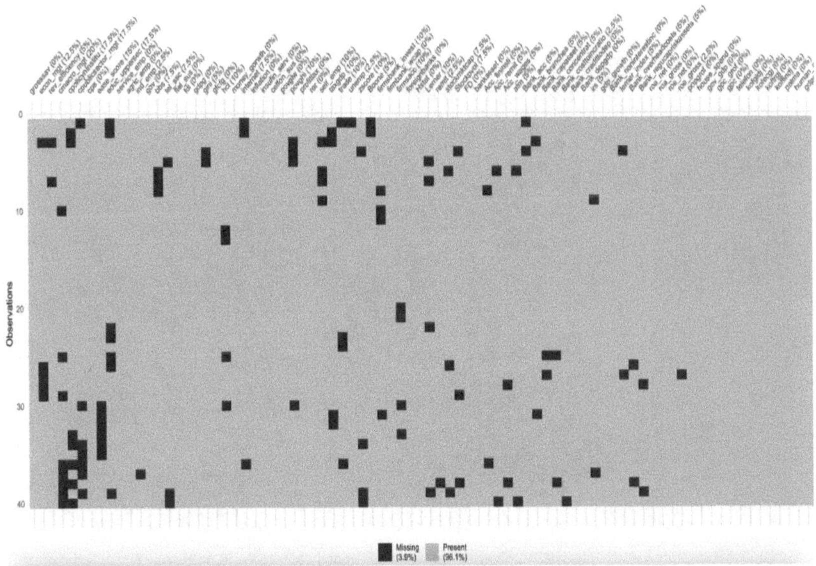

Fig. 2.2 Overview of the dataset before data engineering

method. We opt for the Stratified method since: (1) financial development is negatively skewed, and (2) it ensures a balanced representation in both the testing and training samples.

2.4 Results and Discussion

2.4.1 Exploratory Data Analysis

For brevity, we present the summary statistics, the data partitioning result of financial development across the testing and training samples, and the distribution of financial development at level and log-transformed. Information gleaned from Table 2.3 reveals an average financial development score of 0.103 in the training set as compared to 0.111 in the testing set. Additionally, we observe a mean policy and institutional effectiveness score for the training set is 3.575 as compared to 3.625 of the testing set. Further, we observe a mean per capita income of US$3037.82 and US$3264.43 for the training and testing sets respectively.

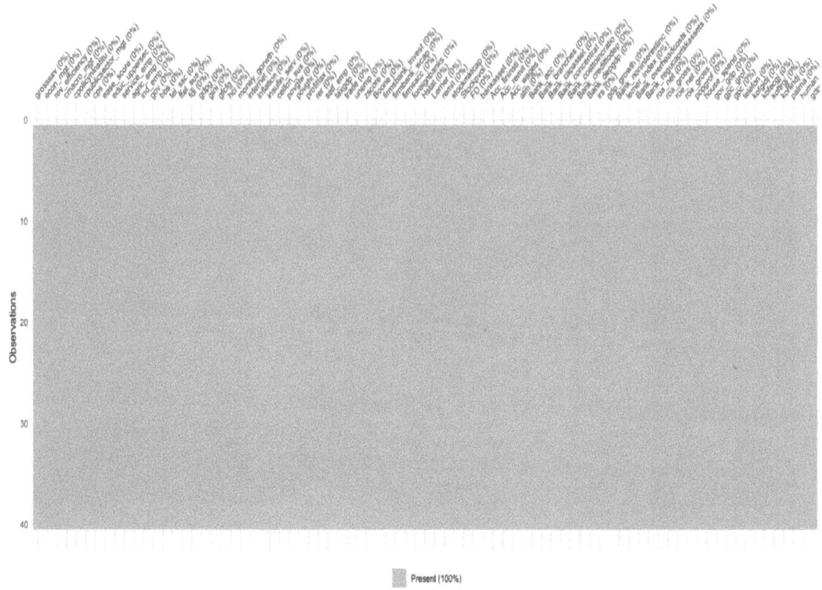

Fig. 2.3 Overview of the dataset after data engineering

2.4.2 Data Partitioning Results

Figure 2.4 shows the 50–50 split of financial development across the two samples. Clearly, the distribution of the data over the training and testing samples follow similar distribution, permitting sound regularisation.

2.4.3 Regularisation Results on Drivers of Financial Development

We now turn attention to the results for Objective 1 of this chapter. As reported in Table 2.4, we find that each shrinkage model—the *Standard lasso, Minimum BIC lasso, and Adaptive lasso* selects different sets of covariates as the main determinants of financial development in Ghana.

We observe from Table 2.4 that, the '*Standard lasso*' algorithms select 17 covariates out of the total 81 as drivers of financial development in Ghana at a penalty (λ) value of 0.0002. Similarly, with a tuning parameter (λ) of 0.0013, the '*Minimum BIC lasso*' estimator selects 12 variables. Finally, we find a more parsimonious shrinkage results in the '*Adaptive*

Table 2.3 Summary statistics for training and testing sets

Variables	*Years*	*Mean (Training set)*	*Mean (Testing set)*	*Std. Dev (Training set)*	*Std. Dev (Testing set)*	*Min (Training set)*	*Min (Testing set)*	*Max (Training set)*	*Max (Testing set)*
Grosssav	40	13.059	13.26	6.408	6.53	4.295	4.006	23.047	23.394
Econ mgt	40	3.975	3.843	0.364	0.508	3	2.5	4.167	4.167
Rev efficiency	40	4.35	4.3	0.328	0.34	3.5	3.5	4.5	4.5
Cfinsector_rating	40	3.5	3.475	0.229	0.302	3	3	4	4
Cmacro_mgt	40	3.85	3.725	0.286	0.525	3	2	4	4
Cpolicy_Institu	40	3.575	3.625	0.183	0.222	3.5	3.5	4	4
Cpublic_sector mgt	40	3.705	3.695	0.06	0.1	3.6	3.4	3.9	3.9
Cps	40	8.489	10.114	5.232	5.178	1.542	1.847	15.658	15.882
Ease_score	40	57.211	57.218	0.717	0.792	56.994	56.994	59.96	60.434
Educ_uppersec	40	7.073	8.759	6.921	7.929	3.7	3.7	20.564	20.564
Service_emp	40	32.711	34.323	6.139	6.754	29.02	29.02	48.959	48.465
Agric_emp	40	52.605	50.742	8.075	8.697	29.265	30.433	56.968	56.968
Ind_emp	40	14.683	14.934	2.16	2.131	13.667	13.881	21.776	21.102
Bbs	40	14,495.45	23,773.8	23,793.132	32,474.607	900	900	68,795	86,596
Tel_ssc	40	160,000	177,000	126,000	115,000	37,129	37,103	356,355	376,509
Ftel_sus	40	0.746	0.788	0.475	0.43	0.292	0.29	1.592	1.64
Fdi	40	7.631e+08	1.240e+09	1.200e+09	1.500e+09	2,400,000	2,000,000	3.255e+09	3.490e+09
Gdpg	40	4.101	5.108	3.782	3.46	−6.924	−3.503	8.143	14.047
Gini	40	39.075	40.08	3.358	2.697	35.3	35.3	43.5	43.5
Gfcfg	40	9.538	14.761	13.864	8.406	−23.75	−2.403	19.596	39.004
Hci	40	0.439	0.439	0.001	0.001	0.439	0.439	0.443	0.443
Money_supply	40	17.559	17.811	9.798	11.322	−12.551	−14.45	26.701	40.08
Internet	40	5.503	7.678	11.681	12.49	0	0	37.884	37.884
Inflation	40	28.801	24.16	26.27	24.158	7.176	7.126	122.875	116.504

(continued)

Table 2.3 (continued)

Variables	*Years*	*Mean (Training set)*	*Mean (Testing set)*	*Std. Dev (Training set)*	*Std. Dev (Testing set)*	*Min (Training set)*	*Min (Testing set)*	*Max (Training set)*	*Max (Testing set)*
Insufin_serv	40	1.693	1.664	1.033	1.095	0.521	0.444	4.188	5.023
Celfon_sus	40	27.033	39.218	45.905	53.662	0	0	134.319	137.517
Povglw	40	11.31	10.97	4.121	5.175	3.4	3.4	17.3	17.3
Povghi	40	26.58	25.445	8.568	10.145	10.9	10.9	37	37
Rer	40	441.164	211.485	861.285	367.828	69.246	64.629	3520.534	1709.006
Self_emp	40	81.599	80.818	2.924	3.173	74.355	74.654	83.695	83.695
Taxgdp	40	13.302	13.69	2.714	2.746	11.437	10.769	21.322	21.752
Trade	40	57.351	68.459	29.135	26.48	6.32	10.079	110.046	116.048
Unemp	40	5.557	6.049	1.621	1.62	4.223	4.157	10.1	10.36
Z-score	40	8.478	8.194	0.558	0.825	6.653	5.776	8.982	8.982
Boone	40	−0.155	−0.152	0.011	0.014	−0.178	−0.175	−0.134	−0.132
Firmsbank_invest	40	16.78	17.04	1.905	2.134	16	16	21.2	21.2
Firmbank_wcap	40	21.985	22.18	1.429	1.601	21.4	21.4	25.3	25.3
Firmsacc	40	85.225	85.8	4.213	4.72	83.5	83.5	95	95
Foreignbanks	40	57.9	58.45	3.11	4.211	50	48	63	63
Hstat	40	0.295	0.272	0.051	0.069	0.16	0.16	0.316	0.316
Lerner	40	0.457	0.448	0.06	0.071	0.257	0.193	0.481	0.481
Remit	40	1.463	2.391	1.628	2.653	0.422	0.442	5.994	10.131
Stockpvol	40	9.117	8.856	0.085	0.696	8.867	6.761	9.145	9.145
FD	40	0.103	0.111	0.021	0.021	0.063	0.075	0.147	0.147
Bankasset	40	58.726	55.084	14.024	5.371	52.103	49.907	100	75.355
Acc_fininsti	40	30.711	30.847	3.958	3.291	29.425	29.425	42.284	42.284
Acc_remit	40	6.541	6.252	1.78	1.293	5.962	5.962	11.746	11.746
Acc_wages	40	6.049	5.844	1.264	0.918	5.639	5.639	9.746	9.746

Variables	*Years*	*Mean (Training set)*	*Mean (Testing set)*	*Std. Dev (Training set)*	*Std. Dev (Testing set)*	*Min (Training set)*	*Min (Testing set)*	*Max (Training set)*	*Max (Testing set)*
ATM	40	4.831	5.217	2.478	2.675	3.844	3.844	11.531	11.531
Bank_acc	40	248.425	284.277	145.601	163.582	178.147	178.147	627.307	627.307
Bank_branches	40	4.006	4.301	1.789	1.765	3.062	3.062	8.597	8.597
Bank_capasset	40	11.908	12.221	1.992	1.8	6.4	6.4	14.717	14.773
Bank_concentrat	40	84.313	76.284	26.688	30.369	35.781	32.521	100	100
Bank_costtoincratio	40	45.817	47.217	7.003	8.924	35	30.535	63.757	63.747
Bank_credittodep	40	53.74	55.132	17.112	16.193	21.401	23.285	73.044	81.702
Irs	40	8.633	9.1	1.716	1.392	2.75	3.5	9.5	9.5
Nim	40	13.353	12.335	1.745	2.251	9.056	8.865	16.807	16.502
Bank_noninterestinc	40	31.775	32.761	3.409	4.208	28.846	25.98	40	41.667
Npl	40	16.223	15.384	2.717	3.749	7.9	6.4	19.6	22.7
Bank_overheadcosts	40	7.582	7.203	0.885	1.278	5.289	3.887	9.271	9.722
Bank_regucapriskasset	40	13.49	13.776	3.307	3.028	9.3	11.1	19.133	18.559
Retun_on_asset(net)	40	5.065	4.702	1.499	1.682	1.369	2.243	7.88	8.705
Retrun_on_equity(gross)	40	7.079	6.584	1.956	2.326	1.875	3.005	10.87	12.436
Return_on_equity(net)	40	49.708	46.062	17.831	21.054	11.766	17.306	75.325	93.821
Retun_on_asset(gross)	40	69.471	64.392	24.062	29.241	16.111	25.102	103.896	134.03
Population_growth(urban	40	4.138	4.067	0.636	0.512	3.019	3.328	5.112	4.925
Population_growth	40	2.623	2.552	0.272	0.24	2.163	2.193	3.009	3.045
Household_expenditure	40	4.126	3.663	4.168	2.754	−0.537	−2.792	17.069	12.191
Gov_gdp	40	9.991	10.494	2.384	1.775	5.861	7.259	15.308	13.789
Profittax	40	1.41	2.46	3.805	3.375	−9.542	−5.987	5.769	11.315
Gpc	40	3037.816	3264.425	959.639	986.86	2344.192	2344.192	5412.628	5194.439
Telefon	40	160,000	177,000	126,000	115,000	37,129	37,103	356,355	376,509

(continued)

Table 2.3 (continued)

Variables	*Years*	*Mean (Training set)*	*Mean (Testing set)*	*Std. Dev (Training set)*	*Std. Dev (Testing set)*	*Min (Training set)*	*Min (Testing set)*	*Max (Training set)*	*Max (Testing set)*
Kofgi	40	47.683	50.763	9.065	8.132	36.359	37.841	61.128	61.128
Kofec	40	39.761	42.765	8.255	6.68	27.904	30.113	50.707	51.448
Koftr	40	38.829	43.429	10.276	8.431	21.538	24.686	53.915	54.282
Koffin	40	40.692	42.102	6.965	5.825	30.668	31.641	51.012	52.848
Kofso	40	30.408	33.812	13.224	13.774	17.581	17.601	51.372	52.732
Palma	40	4.663	5.035	0.947	0.782	3.493	3.493	6.229	6.116
Theil	40	0.537	0.573	0.079	0.065	0.433	0.433	0.638	0.649
Gdpgsqr	40	30.406	37.465	17.044	44.814	0.222	4.745	66.316	197.322
Stockmktcap	40	5.852	6.156	1.38	1.2	3.404	5.261	9.112	8.807

Source Authors' construct (2021)

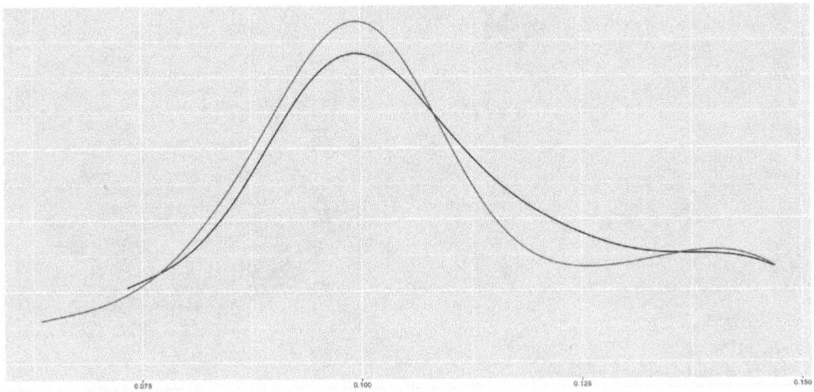

Fig. 2.4 Data partitioning plot, financial development (Stratified method), Training set (Black) and Testing set (Red)

Table 2.4 Determinants of financial development

	Standard_lasso	*Minimum_BIC_lasso*	*Adaptive_lasso*
cpolicynInstitu	X	X	X
human_capital	X	X	X
economic_globalisation	X	X	X
govt_expend	X	X	X
Bank_capital_asset	X	X	X
internet	X		X
gdp_growth	X	X	X
lerner_index	X	X	X
povglow	X	X	
inflation	X		X
unemployment	X		X
real_exchange_rate	X	X	
theil	X		
kofsodj	X		
ease_score	X		
roe_gross	X	X	
Bank_depgdp		X	
moneysupply		X	
_cons	X	X	X

Note X means selected covariate

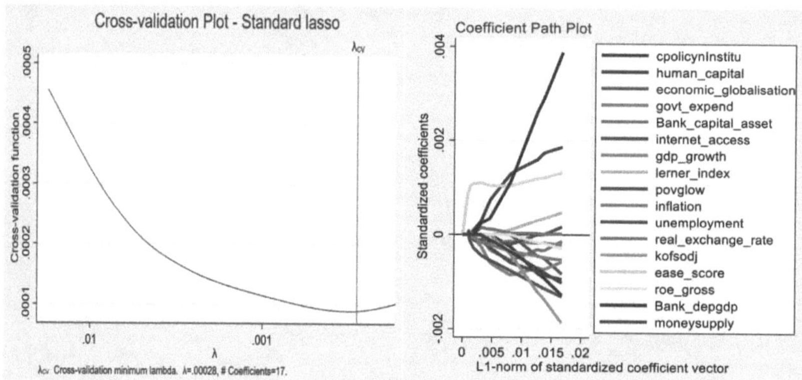

Fig. 2.5 Cross-validation plot (left) and coefficient path plot (right) for Standard lasso

lasso' model as it selects 10 variables out of the 81 covariates. These variables are human capital, internet access, economic growth, economic globalisation, unemployment, inflation, government expenditure, market power (Lerner index), bank asset and institutional quality. The reliability of our results is seen in the post estimation results of cross-validation and coefficient path plots (right-hand side of Figs. 2.5, 2.6, and 2.7), showing that the selected covariates across the three models are well-behaved.

2.4.4 *Pillars of Financial Development in Ghana*

Taking cues from the theoretical and empirical survey, we classify the drivers of Ghana's financial sector development selected by the Adaptive lasso under three pillars—the macroeconomy pillar, the education and ICT pillar, and bank-specific and regulatory pillar (see, Fig. 2.8). To provide inference on these variables, we invoke the lasso linear inferential techniques whose estimates are presented next.

2.4.5 *Lasso Inferential Results*

In this section, estimates from the DSL, POLR and POIVLR are presented. For consistency, we present our estimates under the 3 pillars presented in Sect. 2.4.4 (Table 2.5).

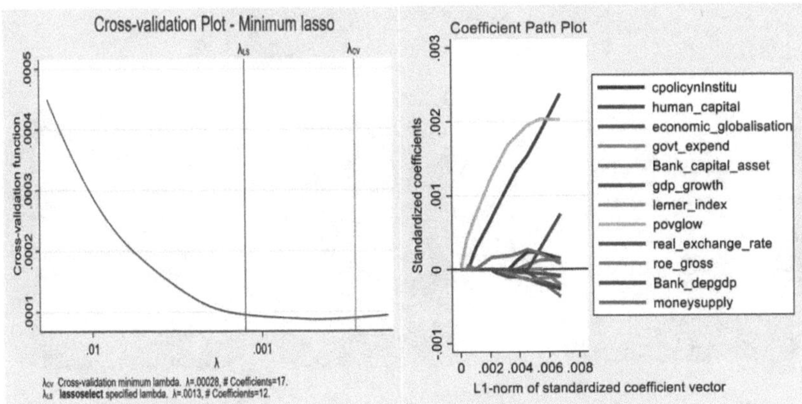

Fig. 2.6 Cross-validation plot (left), and coefficient path plot (right) for Minimum BIC lasso

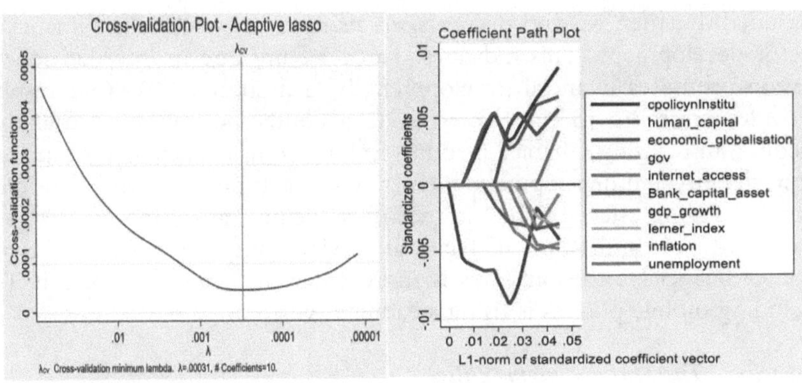

Fig. 2.7 Cross-validation plot (left), and coefficient path plot (right) for Adaptive lasso

2.4.5.1 *The Education and ICT Pillar*

First, albeit statistically insignificant, the results suggest that human capital development is positively related to financial development in Ghana. This stems from the argument that in societies where educational attainment is high, financial products and services are consumed by a significant section of the population. This implies that building a robust financial sector

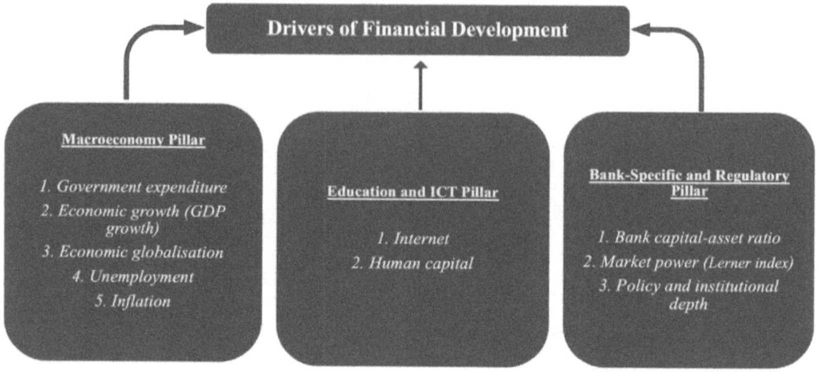

Fig. 2.8 Drivers of financial development based on Adaptive lasso selection

in Ghana would rest on a comprehensive financial capability and education strategy that seek to enlighten the populace especially the financially excluded. Further, we find that *internet access* is key for Ghana's financial sector development. This indicates that a 1% upsurge in internet penetration promotes financial development by a modest 0.001. Our result provides optimism in the use of ICT, a contemporary youth-friendly medium in realising Ghana's medium to long-term financial sector inclusion, stability and development agenda. The result points to the plausible game-changing effect of the internet in reducing transaction costs associated with the utilisation of financial services, and the achievement of greater financial inclusion. This is more so as the usage of smart ICTs including mobile phones is rising steadily in Ghana.

2.4.5.2 *The Macroeconomy Pillar*

On the macroeconomy front, the results show that economic globalisation matters for Ghana's financial sector development. The magnitude of the coefficient indicates that opening up Ghana's economy to the global market enhances the development of the Ghanaian financial sector by 0.002. There are prospects, therefore, that opening up Ghana's economy to both trade, foreign direct investment and capital flows can boost financial development plausibly through innovation, dynamism and competition. However, this must be embraced with cautious optimism as greater financial and economic integration presents the financial sector with greater financial risk in the form of cybercrime, illicit financing and

Table 2.5 Lasso inferential results on the drivers of financial developments in Ghana

Variables	*(1)* *DSL*	*(2)* *POLR*	*(3)* *POIVLR*
Human capital	0.054	0.003	0.003
	(0.034)	(0.034)	(0.034)
Government expenditure	0.001***	0.001***	0.001***
	(0.000)	(0.000)	(0.000)
GDP growth	−0.001	−0.001	−0.001
	(0.000)	(0.000)	(0.000)
Economic globalisation	0.002***	0.002***	0.002***
	(0.000)	(0.000)	(0.000)
Unemployment	−0.001	−0.001**	−0.001**
	(0.000)	(0.000)	(0.000)
Inflation	−0.001***	−0.001***	−0.001***
	(0.000)	(0.000)	(0.000)
Internet access	0.001***	0.001***	0.001***
	(0.000)	(0.000)	(0.000)
Institutions	0.012**	0.015***	0.015***
	(0.005)	(0.004)	(0.004)
Bank assets	0.001**	0.001***	0.001***
	(0.000)	(0.000)	(0.000)
Lerner index	0.015***	0.016**	0.016**
	(0.006)	(0.007)	(0.007)
Observations	1,722	1,722	1,722
Wald	250.6	564.8	564.8
ChiP-value	0.000	0.000	0.000
Vars of Interest	10	10	10
Controls	48	48	48
Controls selected	45	45	45
Instruments	–	–	10

Robust Standard errors in parentheses
***$p < 0.01$, **$p < 0.05$, *$p < 0.1$

greater susceptibility to global financial crisis. Additionally, we find that government expenditure and economic growth drive Ghana's financial sector development. While we do not find empirical evidence for the latter, the former reports a 0.001 statistically significant effect on Ghana's financial sector. Indeed, as a developing country, the consumption of financial products and services by the government can boost financial sector performance. However, excessive borrowing can also crowd out private investment or result in inefficient resource allocation. Therefore, we are of the view that the financial sector is likely to develop if expenditure results in a more lubricated economy, plausibly through the provision of resources to (1) develop the payment system platforms and services; (2) support innovation and efficient delivery of financial services while enhancing information flow to consumers[9]; (3) strengthen the secured transactions and collateral frameworks and the insolvency regime. This is also linked to strategies and policies aimed at reducing unemployment, which can boost financial development through investment, and reduction in non-performing loans. Finally, we find that maintaining stable inflation is equally germane for Ghana's financial sector development. The result reveals that financial development reduces by 0.001 if the inflation rate rises by 1%. Indeed, as an inflation-targeting economy, high inflation rates can be a disincentive to savings and investment plans. Further, rising inflation levels can result in exchange rate volatility, which can have harmful ramifications for arbitrageur and investor plans and decisions. Therefore, to boost consumer confidence in the utilisation of the financial system for transactions, maintaining stable inflation is worthwhile.

2.4.5.3 *The Bank-Specific and Regulatory Pillar*

Our evidence shows that financial sector strength, proxied by bank capital to asset ratio and market power (i.e., Lerner index) also matter for Ghana's financial sector development. In specifics, we report a favourable financial development effect of 0.001 for the former and approximately 0.02 for the latter. Additionally, we provide strong empirical evidence to show that regulatory quality promotes financial development. This is also linked with the internet and ICT pillar, which signifies that to ensure continuous utilisation of online banking services, building consumer confidence is crucial. This is more so considering the implementation of the African Continental Free Trade Area agreement, which is expected to enhance economic and financial globalisation.

2.4.6 *Modelling Financial Sector Development in Ghana*

We now turn to the results on the second objective of the chapter, which is the presentation of a model best for predicting financial development

[9] Tightening the national identification and tax identification systems.

Table 2.6 Evaluation of lasso models

Regularisation models	*Sample*	*MSE*	*R-squared*
Standard lasso	*Training sample*	0.004	0.880
	Testing sample	0.103	0.728
Minimum BIC lasso	*Training sample*	0.005	0.868
	Testing sample	0.103	0.726
Adaptive lasso	*Training sample*	0.004	0.880
	Testing sample	**0.102**	**0.728**

Note MSE is Mean Squared Error
Source Authors' construct (2021)

in Ghana. To this end, we evaluate the prediction accuracy (i.e., the goodness-of-fit statistics) of the three shrinkage models. The results as presented in Table 2.6 shows that the best model is the Adaptive lasso.

The appropriateness of the Adaptive lasso as the best model for predicting financial development lies in its relative lower out of sample error (i.e., MSE) of 0.102, and R-squared of 72.8%. Consequently, we model Ghana's financial development as:

$$\begin{aligned} fd_t = \beta_0 + \beta_1\, institutions_t + \beta_2\, growth_t + \beta_3\, humuancap_t \\ + \beta_4\, glob_t + \beta_5\, internet_t + \beta_6\, gov_t + \beta_7\, bankasset_t \\ + \beta_8\, marketpower_t + \beta_9\, inf_t + \beta_{10}\, unempl_t + \varepsilon_t \end{aligned} \tag{2.8}$$

where ***fd*** is financial development; ***institution*** is financial sector regulatory effectiveness score; ***growth*** is GDP growth; ***humancap*** is human capital development; ***glob*** is economic globalisation; ***internet*** is internet access; ***gov*** is government expenditure; ***bankasset*** is bank capital to asset ratio; ***inf*** is inflation; and ***unempl*** is unemployment.

2.5 Conclusion

This chapter employs machine learning techniques to identify (1) the key determinants of financial development in Ghana, and (2) a model best for predicting Ghana's financial sector development. To this end, we employ a large dataset comprising 81 potential determinants of financial development over the period 1980–2019 for the analysis. On the first objective, evidence from the Adaptive lasso shows that out of the 81 possible predictors of financial sector development in Ghana, only *human capital*, *internet access, economic growth, economic globalisation, unemployment, inflation, government expenditure, market power, bank asset and regulatory quality* are key. The chapter, therefore, confirms the demand-leading hypothesis, growth-led theory and inflation and finance theory.

For policy recommendations, we suggest that the Ghanaian government invests strategically in human capital development, the new economy and the productive capacity of the economy. This calls for government intervention in providing affordable and reliable internet access, especially for rural dwellers who are more likely to use mobile money services. It also calls for financial institutions and markets providing services online while incentivising clients using such platforms through discounts. It, therefore, requires a task force instituted by the BoG to oversee online transactions and the activities of forex bureaux. Further, we recommend that policymakers strengthen Ghana's financial sector regulatory regime to enhance consumer confidence and protection, information flow and financial sustainability. Another point of caution is that to avoid a heating-up of the economy, the pace of financial sector growth should move in tandem with the growth of the economy, particularly, following the implementation of the AfCFTA and the projected rise in financial globalisation in Africa. In this regard, concerted effort by the monetary and fiscal authorities is needed for building robust frameworks for boosting financial system information flow, competition and the reduction in financial risk. This can be achieved through the revision of the Bank of Ghana's prudential standards as well as enhancement in on-site and off-site supervision. It also requires effective collaboration with institutions such as the African Development Bank, the Organization for Economic Co-operation and Development and the World Bank in strengthening the oversight capacity of the Bank of Ghana to deal with issues of money laundering and illicit financing characteristic of enhanced globalisation.

References

Acemoglu, D., Johnson, S., & Robinson, J. A. (2001). The colonial origins of comparative development: An empirical investigation. *American Economic Review, 91*(5), 1369–1401.

Adu, G., Marbuah, G., & Mensah, J. T. (2013). Financial development and economic growth in Ghana: Does the measure of financial development matter? *Review of Development Finance, 3*(4), 192–203.

Aggarwal, R., Demirgu¨ç-Kunt, A., & Pería, M. S. M. (2011). Do remittances promote financial development. *Journal of Development Economics, 96*(2), 255–264.

Ahmed, H. (2013). Financial inclusion and Islamic finance: Organizational formats, products, outreach, and sustainability. *Economic Development and Islamic Finance*, 203–230.

Almarzoqi, R., Naceur, M. S. B., & Kotak, A. (2015). *What matters for financial development and stability?* (No. 15–173). International Monetary Fund.

Aluko, O. A., & Ajayi, M. A. (2018). Determinants of banking sector development: Evidence from Sub-Saharan African countries. *Borsa Istanbul Review, 18*(2), 122–139.

Aluko, O. A., & Ibrahim, M. (2019). On the macroeconomic determinants of financial institutions development in sub-Saharan Africa. *International Review of Economics*, 1–17.

Aluko, O. A., & Ibrahim, M. (2020). Institutions and financial development in ECOWAS. *Journal of Sustainable Finance & Investment*, 1–12.

Arcand, J., Berkes, E., & Panizza, U. (2012). *Too much finance?* (IMF Working paper WP/12/161). International Monetary Fund.

Aryeetey, E., Nissanke, M., & Steel, W. F. (2000). Intervention and liberalization: Changing policies and performance in the financial sector. In E. Aryeetey, J. Harrigane, & M. Nissanke (Eds.), *Economic reforms in Ghana: The miracle and the Mirage*. James Currey and Woeli Publishers.

Asongu, S. A. (2012). Bank efficiency and openness in Africa: Do income levels matter? *The Review of Finance and Banking, 4*(2), 115–122.

Ayadi, R., Arbak, E., Naceur, S. B., & De Groen, W. P. (2015). Determinants of financial development across the Mediterranean. In *Economic and social development of the southern and eastern Mediterranean countries* (pp. 159–181). Springer.

Baltagi, B. H., Demetriades, P. O., & Law, S. H. (2009). Financial development and openness: Evidence from panel data. *Journal of Development Economics, 89*(2), 285–296.

Bandura, W. N. (2020). Trade openness, institutions and financial development in Sub-Saharan Africa. *Studies in Economics and Econometrics, 44*(2), 29–48.

Barajas, M. A., Beck, T., Dabla-Norris, M. E., & Yousefi, M. R. (2013). *Too cold, too hot, or just right? Assessing financial sector development across the globe* (No. 13–81). International Monetary Fund.

Beck, T., Demirgu¨ç-Kunt, A., & Levine, R. (2003). Law, endowments and finance. *Journal of Financial Economics, 70*(2), 137–181.

Beck, T., & Levine, R. (2005). Legal institutions and financial development. In *Handbook of new institutional economics* (pp. 251–278). Springer.

Belloni, A., Chen, D., Chernozhukov, V., & Hansen, C. (2012). Sparse models and methods for optimal instruments with an application to eminent domain. *Econometrica, 80*(6), 2369–2429.

Belloni, A., Chernozhukov, V., & Hansen, C. (2014). High-dimensional methods and inference on structural and treatment effects. *Journal of Economic Perspectives, 28*(2), 29–50.

Bittencourt, M. (2011). Inflation and financial development: Evidence from Brazil. *Economic Modelling, 28*(1–2), 91–99.

Boopen, S., Kesseven, P., Jashveer, H., & Binesh, S. (2011). *Determinants of financial development: The case of Mauritius*. Finance and Corporate Governance Conference.

Boyd, J. H., Levine, R., & Smith, B. D. (2001). The impact of inflation on financial sector performance. *Journal of Monetary Economics, 47*(2), 221–248.

Cherif, M., & Dreger, C. (2016). Institutional determinants of financial development in MENA countries. *Review of Development Economics, 20*(3), 670–680.

Chernozhukov, V., Hansen, C., & Spindler, M. (2015). Valid post-selection and post-regularization inference: An elementary, general approach. *Annual Review of Economics, 7*(1), 649–688.

Čihák, M., Demirgüč-Kunt, A., Feyen, E., & Levine, R. (2013). *Financial development in 205 economies, 1960 to 2010 (No. w18946)*. National Bureau of Economic Research.

Cooray, A. (2011). The role of the government in financial sector development. *Economic Modelling, 28*(3), 928–938.

Coulibaly, D. (2015). Remittances and financial development in Sub-Saharan African countries: A system approach. *Economic Modelling, 45*, 249–258.

David, M. A., Mlachila, M. M., & Moheeput, A. (2014). *Does openness matter for financial development in Africa?* International Monetary Fund.

Demetriades, P., & Hook Law, S. (2006). Finance, institutions and economic development. *International Journal of Finance & Economics, 11*(3), 245–260.

Djankov, S., McLiesh, C., & Shleifer, A. (2007). Private credit in 129 countries. *Journal of Financial Economics, 84*(2), 299–329.

Elsherif, M. A. (2015). The determinants of financial development: Empirical evidence from Egypt. *The Macrotheme Review, 4*(3), 69–87.

Feyen, E., Katie K., & Diego S. (2013). *Fin Stats 2014: A ready-to-use tool to benchmark financial sectors across countries and over time*. World Bank Financial Systems Global Practice.

Filippidis, I., & Katrakilidis, C. (2014). Institutions, policy and banking sector development: A reassessment. *Finance a Uver, 64*(6), 501.

Fowowe, B. (2014). Law and finance revisited: Evidence from African countries. *South African Journal of Economics, 82*(2), 193–208.

Government of Ghana. (2018). *National financial inclusion and development strategy, 2018–2023*. Ministry of Finance, Accra, Ghana.

Gygli, S., Haelg, F., Potrafke, N., & Sturm, J. E. (2019). The KOF globalisation index—Revisited. *Review of International Organizations, 14*(3), 543–574.

Hastie, T., Tibshirani, R., & Friedman, J. (2001). *The elements of statistical learning: Data mining, inference, and prediction*. Springer Series in Statistics.

Hastie, T., Tibshirani, R., Friedman, J. H., & Friedman, J. H. (2009). *The elements of statistical learning: Data mining, inference, and prediction* (Vol. 2, pp. 1–758). Springer.

Huybens, E., & Smith, B. D. (1999). Inflation, financial markets and long-run real activity. *Journal of Monetary Economics, 43*(2), 283–315.

Ibrahim, M., & Sare, Y. A. (2018). Determinants of financial development in Africa: How robust is the interactive effect of trade openness and human capital? *Economic Analysis and Policy, 60*, 18–26.

James, G., Witten, D., Hastie, T., & Tibshirani, R. (2013). Linear model selection and regularization. In *An introduction to statistical learning* (pp. 203–264). Springer.

Khalfaoui, H. (2015). The determinants of financial development: Empirical evidence from developed and developing countries. *Applied Economics and Finance, 2*(4), 1–9.

Kodila-Tedika, O., & Asongu, S. A. (2015). The effect of intelligence on financial development: A cross-country comparison. *Intelligence, 51*, 1–9.

Lahoti, R., Jayadev, A., & Reddy, S. (2016). The global consumption and income project (GCIP): An overview. *Journal of Globalization and Development, 7*(1), 61–108.

La Porta, R., Lopez-de-Silanes, F., Shleifer, A., & Vishny, R. W. (1998). Law and finance. *Journal of political economy, 106*(6), 1113–1155.

Law, S., & Demetriades, P. O. (2006). *Openness, institutions and financial development* (WEFRP Working Paper No. WEF 0012).

Liang, Q., & Jian-Zhou, T. (2006). Financial development and economic growth: Evidence from China. *China Economic Review, 17*(4), 395–411.

Mahawiya, S. (2015). Financial sector development, inflation and openness: A comparative panel study of ECOWAS and SADC. *Economic Research Southern Africa Working Paper, 528*, 1–37.

Marcelin, I., & Mathur, I. (2014). Financial development, institutions and banks. *International Review of Financial Analysis, 31*, 25–33.

Mavrotas, G., & Santillana, M. (1999). *Savings and financial sector development: Key issues*. International Conference on Finance and Development, University of Manchester, 9–10.

McKinnon, R. I. (1973). *Money and capital in economic development*. Brookings Institution.

McKinnon, R. I. (1993). *The order of economic liberalization: Financial control in the transition to a market economy*. JHU Press.
Mtar, K., & Belazreg, W. (2020). Causal nexus between innovation, financial development, and economic growth: The case of OECD countries. *Journal of the Knowledge Economy*, 1–32.
Naceur, S. B., Cherif, M., & Kandil, M. (2014). What drives the development of the MENA financial sector? *Borsa Istanbul Review, 14*(4), 212–223.
Naqvi, T., Waheed, A., Mahmood, H., & Rafique, M. (2017). Impact of political instability on financial development of Pakistan. *International Journal of Management Sciences and Business Research, 6*(4).
Nguyen, C. P., Su, T. D., & Doytch, N. (2020). The drivers of financial development: Global evidence from internet and mobile usage. *Information Economics and Policy, 53*, 100892.
Opperman, P., & Adjasi, C. K. D. (2019). Remittance volatility and financial sector development in sub-Saharan African countries. *Journal of Policy Modeling, 41*(2), 336–351.
Padachi, K., Rojid, S., & Seetanah, B. (2008). Investigating into the factors that influence the adoption of internet banking in Mauritius. *Journal of Internet Business* (5).
Peprah, J. A., Ofori, I. K., & Asomani, A. N. (2019). Financial development, remittances and economic growth: A threshold analysis. *Cogent Economics & Finance, 7*(1), 1625107.
Raza, S. H., Shahzadi, H., & Akram, M. (2014). Exploring the determinants of financial development (using panel data on developed and developing countries). *Journal of Finance and Economics, 2*(5), 166–172.
Robinson, J. C. (1952). *The generalization of the general theory. In The rate of interest and other essays*. Macmillan.
Rousseau, P. L., & Wachtel, P. (2002). Inflation thresholds and the finance-growth nexus. *Journal of International Money and Finance, 21*, 777–795.
Sanusi, K. A., Meyer, D., & Ślusarczyk, B. (2017). The relationship between changes in inflation and financial development. *Polish Journal of Management Studies, 16*(2), 253–265.
Schneider, U., & Wagner, M. (2012). Catching growth determinants with the adaptive lasso. *German Economic Review, 13*(1), 71–85.
Schwarz, G. (1978). Estimating the dimension of a model. *Annals of Statistics, 6*(2), 461–464.
Shabbir, B., Jamil, L., Bashir, S., Aslam, N., & Hussain, M. (2018, February 13). *Determinants of financial development. A case study of Pakistan*. A Case Study of Pakistan.
Shaw, E. S. (1973). *Financial deepening and economic development*. Oxford University Press.

Svirydzenka, K. (2016). *Introducing a new broad-based index of financial development* (IMF Working Paper No. 16/5).

Takyi, P. O., & Obeng, C. K. (2013). Determinants of financial development in Ghana. *International Journal of Development and Sustainability, 2*(4), 2324–2336.

Tayssir, O., & Feryel, O. (2018). Does central banking promote financial development? *Borsa Istanbul Review, 18*(1), 52–75.

Tibshirani, R. (1996). Regression shrinkage and selection via the lasso. *Journal of the Royal Statistical Society: Series B (Methodological), 58*(1), 267–288.

Tibshirani, R. J., & Taylor, J. (2012). Degrees of freedom in lasso problems. *The Annals of Statistics, 40*(2), 1198–1232.

Voghouei, H., Azali, M., & Jamali, M. A. (2011). A survey of the determinants of financial development. *Asian-Pacific Economic Literature, 25*(2), 1–20.

World Bank. (2021). *World development indicators*. World Bank.

World Bank. (2019). *Global financial development report 2019/2020: Bank regulation and supervision a decade after the global financial crisis*. World Bank.

Zou, H. (2006). The adaptive lasso and its oracle properties. *Journal of the American Statistical Association, 101*(476), 1418–1429.

Zou, H., Hastie, T., & Tibshirani, R. (2007). On the "degrees of freedom" of the lasso. *The Annals of Statistics, 35*(5), 2173–2192.

CHAPTER 3

Harnessing Financial Innovation for Financial Inclusion in Ghana

Joshua Sebu, Eric Amoo Bondzie, Ewura-Adwoa Ewusie, and Jonathan Tawiah-Mensah

3.1 Introduction

In recent years, financial innovation has become the catalyst underpinning the transformation of financial services in advanced economies and is rapidly altering the modus operandi of financial services in emerging markets and middle-income countries. Recent evidence suggests that various new initiatives[1] such as the use of Faster Payment Systems (FPS)

[1] Project Inthanon, Faster Payment System (FPS), Portugal Finlab, Project Jasper, among others.

J. Sebu (✉)
Department of Data Science and Economic Policy, University of Cape Coast, Cape Coast, Ghana
e-mail: Joshua.sebu@ucc.edu.gh

E. A. Bondzie · E.-A. Ewusie
Department of Economic Studies, University of Cape Coast, Cape Coast, Ghana
e-mail: eric.bondzie@ucc.edu.gh

J. A. Peprah et al. (eds.), *Financial Sector Development in Ghana*, Palgrave Macmillan Studies in Banking and Financial Institutions,
https://doi.org/10.1007/978-3-031-09345-6_3

have been introduced by some central banks to expedite the transfer of funds and the conduct of financial service delivery (Bank of Lithuania, 2019; Bank of Thailand, 2019; Danske Bank Group, 2007; Hong Kong Monetary Authority [HKMA], 2018). Joint initiatives are also being pursued among countries to advance the frontiers of the institutional and regulatory framework. The partnership between Canada and Portugal seeks to promote the crucial Distributed Ledger Technology (DLT)[2] and implement a more accessible financial regulatory environment (Bank of Portugal, 2018; Chapman et al., 2017). Such novel developments are promoting and strengthening financial services delivery in these countries and around the world.

Financial innovation simply refers to the use of technological advances[3] to facilitate financial transactions and can be broadly classified into process, product and institutional innovation. Process innovation identifies new ways of operating a business and implementing information technology, such as ATMs, mobile banking and online banking (Abor, 2005). Product innovation includes new financial products such as securitised assets, derivatives and retail structured products; and lastly, institutional innovation is a process of introducing new types of financial firms such as discount broking firms, internet banks and specialist credit card firms (Tahir et al., 2018). Avlonitis et al. (2001) summarises financial innovation as the improvements in development processes in terms of activities, formality, cross-functional involvement and performance outcomes. The continual departure from existing methods of operation is shifting the nature of financial intermediation, financial markets activities and the overall financial infrastructure within which the financial system operates.

As observed around the world, the relentless wave of financial innovation is also rippling through some African countries with great success.

E.-A. Ewusie
e-mail: ewura.ewusie@ucc.edu.gh

J. Tawiah-Mensah
UKRI-GCRF One Ocean Hub, University of Cape Coast, Cape Coast, Ghana

[2] A secure storage protocol fostering exceptional levels of innovation.

[3] The word financial innovation is limited to new or altered financial instruments (i.e. deposit and cash management instruments) issued by banks and other deposit-taking institutions. The term also means the inclusion of new financial instruments in financial institutions and markets through new technologies.

The South African Reserve Bank recently released a report on an inter-bank wholesale settlement using tokenised South Africa Rand on DLT, which has significantly reduced the time needed to process a typical daily volume of payments (South African Reserve Bank, 2018). In Kenya, Safaricom's M-Pesa, which was launched in 2007, is acclaimed as the world's successful money transfer service (Vodafone, 2019). This service has enabled scores of people who own a mobile devise but have a limited access to, or do not own a bank account to conduct vital financial transactions on their phones (Vodafone, 2019). Financial innovation is, therefore, aimed at the efficient delivery of products and services to improve customer satisfaction and business development.

Recently, Ghana has embarked on a similar drive to improve the efficiency of its financial sector by resorting to new financial products and services. For instance, the Central Bank of Ghana revealed that the range of payment service providers has expanded to include mobile money operators and financial technology—FinTech firms, which work closely with the major financial institutions to develop innovative and affordable products for the unbanked population in Ghana (Bank of Ghana, 2018). These products include online payment platforms, unstructured supplementary service, data services, quick response codes, money transfer services, deployment of point of sale (POS) devices and agency banking. The bank has also implemented a wide range of activities such as Digital Agent Registry, payment of interest on e-money float balances, cybersecurity regulations and mobile money interoperability project to further deepen the activities of the network.

According to Nkem and Akujinma (2017), the banking sector in most of the developing countries has gained strength due to financial innovations in various payment systems. This demonstrates that financial innovation has eased information and transaction costs and made the use of financial services much more accessible. With the majority of economic activity pursued in the informal sector, this transformation has become even more paramount for financial inclusion in developing countries. Although financial innovation is largely progressive, there are also some emerging threats such as online fraud and card cloning which pose real challenges for the sector and may need to be addressed if the financial innovation strategy would succeed in improving the overall efficiency of the sector and ensure customer confidence in the financial system.

Despite the relentless penetration of financial innovation in Ghana, very little is documented on the innovation efforts and emerging challenges. Hence, to gain a broader perspective of the overall financial landscape in the country, it is pertinent that the innovation paraphernalia is explored to provide information on this vital area. Against this backdrop, this chapter outlines the nature and extent of financial innovation in Ghana. Our goal is to describe adequately, the projects, processes and products being harnessed to improve financial inclusion given that a greater proportion of the population is unbanked. Hence, we do not present correlates or determinants of financial innovation. Since efforts at harnessing financial innovation are currently being pursued by different institutions, this chapter synthesises initiatives by the Central Bank and Ministry of Finance—and the traditional and emerging financial players such as the banks and telecommunication companies. The information is complemented by an analysis of some key stakeholders on the financial inclusion programme, including its benefits and challenges.

The rest of this chapter is structured as follows. Section 3.2 submits an overview of financial innovation tools in some advanced, emerging and developing countries as a background to the innovation initiatives being developed in Ghana. Section 3.3 presents the methodology. Section 3.4 provides the various initiatives which depict the nature and extent of financial innovation in Ghana. Section 3.5 analyses stakeholder perspectives on the effort to ensure financial inclusion as well as its merits and emerging challenges. Section 3.6 presents the conclusion.

3.2 Overview of Financial Innovation Tools

Generally, the concept of financial innovation embraces changes in the products and services offered by banks, insurance companies, investment funds and other financial service firms as well as modifications to internal structures and processes, managerial practices and new ways of interacting with customers (Mention & Torkkeli, 2012). Globally, the trend of financial service delivery is shifting from the paper-based system to the technology-based system. The shift has paved the way for the intense use of technology (Rose, Marquis and Lu, 2009). Such innovations in the financial sector are fast-growing to the extent that a lot of financial houses around the world see the future of financing as being led by more sophisticated robotics tools (PricewaterCooper, 2016). This revolution has contributed to what is now known as FinTech[4] in the global financial space; a movement constantly shaping and rewriting the trajectory of financial sector development.

FinTech describes *'incremental or disruptive innovations in the context of the financial services industry as induced by information technology (IT) developments resulting in new intra or inter-organizational business models, products and services, organizations, processes, and systems'* (Child et al., 2019). The financial technology-based operations include, but are not limited to, online marketplace lending (called peer-to-peer lending), equity crowdfunding, Robo-advice, financial applications of DLT and financial applications of machine learning commonly referred to as artificial intelligence and machine intelligence (Schindler, 2017). FinTech has now become the platform accessed and utilised by several countries and stakeholders around the world; thus, benefiting lots of financial institutions and their clients.

FinTech is targeted at helping companies, business owners and consumers to manage their financial processes and lives efficiently by utilising specialised software and algorithms that are used on computers and, increasingly, on smartphones. The Fintech services range from payment solutions and information services, savings and deposit-taking, online banking, advisory services, securities trading services, simple financing solutions and other modern financial software. Services like mobile payment systems hold the promise of lucrative market entry and growth opportunities for many players outside the banking sector (Deutsche Bank, 2014). Hence, the evolution of financial technology and the rapid emergence of new innovations have fundamentally changed how securities and payments are cleared and settled in many parts of the world.

Project Jasper was launched in March 2016 as a collaborative experimental initiative by Payments Canada, the Bank of Canada, financial innovation consortium R3 and a number of Canadian financial institutions to understand how DLT could transform the future of payments. The immediate goal of Project Jasper was to build a proof-of-concept system (with no intention of advancing to a production-level system) that leveraged a settlement asset issued and controlled by a central bank. The Project Jasper provided vital insights into how a central bank and

[4] FinTech stands for 'Financial Technologies', and it encompasses all technologies used and applied in the financial services sector.

participating financial institutions can complete interbank payments on a distributed ledger (Chapman et al., 2017).

In Japan, Seven Bank has established a tie-up for ATM usage that will enable deposits and withdrawals using the LINE Pay app, a smartphone wallet service provided by LINE Pay Corporation, at Seven Bank ATMs. LINE Pay is a mobile payment platform provided through LINE, a communication app with more than 64 million monthly users in Japan. It allows users to simply and conveniently remit funds to other users and make payments to participating services and stores, within the limit of the funds deposited in their LINE Pay accounts. In addition, for members who want a card, the LINE Pay Card is issued for use at approximately 31 million JCB affiliates worldwide (Seven Bank, 2016).[5]

In many emerging and developing countries, financial innovation has contributed to the growth of the financial sector and has also become an essential part of the economy. For instance, the Chinese financial market is more conducive to accepting QR codes for mobile payments. QR codes tend to dominate in markets where merchants need a mobile phone infrastructure but have limited resources to support Near-Field Communications Point-of-Sale (NFC POS) upgrades.[6] QR codes provide an opportunity for a large number of shops to print merchant-presented static QR codes that consumers scan to enable merchants to accept mobile payments (Pandy & Crowe, 2019).[7] For instance, Alipay provides domestic and cross-border payment services through partnerships with more than 200 financial institutions including leading national and regional banks across China, as well as Visa and MasterCard. As at 2013, there were 300 million real-name registered users on Alipay, 250 million active monthly users, 100 million plus daily transactions, US $600 billion

[5] JCB (formerly Japan Credit Bureau) is the only international payment brand based in Japan. It offers high quality payment solution and technologies that are trusted worldwide since1981. JCB cards are also accepted in Australia, New Zealand and Canada with the partnership with American Express, and in the U.S. through Discover Network. It has also partnered with UnionPay enabling cards from both brands to be used at selected merchants in Japan and China.

[6] NFC POS is the same technology used by Apple Pay, Samsung Pay, Android Pay and others. This enables a payment card or smartphone to 'talk' to a nearby in-store POS terminal as long as that terminal is also NFC-enabled.

[7] A typical example is Alipay. Alipay is a QR code-based mobile wallet; an electronic payment system and the leading third-party online payment platform with the biggest market share in China.

transaction volume and 48.7% share of third-party online payment market (Alipay, 2015).

The Hong Kong Monetary Authority (HKMA) announced seven initiatives in September 2017 to prepare Hong Kong to move into the era of smart banking. One of these initiatives is the Faster Payment System (FPS) which operate on a round-the-clock basis and connects banks and stored value facility (SVF) operators on the same platform. It enables the public to transfer funds anytime, anywhere, across different banks or SVFs with funds available almost immediately. Fund transfer becomes user-friendly with the use of mobile number or email address as an account proxy for the payee (HKMA, 2018).

The South African Reserve Bank (SARB) Fintech Unit has released its first project named 'Khokha' with the goal to contribute to the global initiatives which assess the application and use cases of Distributed Ledger Technology (DLT) through this collaborative effort. Formed in August 2017, Project Khokha builds on the initiatives undertaken by central banks in Brazil, Canada, some countries in Europe, Japan and Singapore. The DLT nodes of Project Khokha is operated under a variety of deployment models (on-premise virtual machine and cloud) and across distributed sites while processing the current South African real-time gross settlement system's high-value payments transaction volumes within a two-hour window (South African Reserve Bank, 2018).

M-Pesa in Kenyan is the world's most successful money transfer service. It enables millions of people who have access to a mobile phone, but do not have or have only limited access to a bank account, to send and receive money, top-up airtime, make bill payments and much more. M-Pesa was launched by Vodafone's Kenyan associate, Safaricom, on 6 March 2007 (Vodafone, 2019). Another major change came in 2016 when Visa launched a mobile payment application in Kenya as a challenge to incumbent M-Pesa. It partnered four large banks to launch mVisa which enables consumers to make payments by scanning a merchant's QR code. Upon its success, the Visa initiative is currently being implemented in some developing countries including Ghana (Vodafone, 2019).

3.3 Methodology

This chapter adopted a qualitative exploratory approach to describe the nature and scope of financial innovation in Ghana. It relied solely on secondary data extracted from reports from the central bank, commercial/universal banks, telecommunication companies. It also utilised extracts from media interview[8] reports that included the voices of the Governor of the Bank of Ghana, the first deputy Governor of the Bank of Ghana, and a senior financial sector specialist and Co-author of the 2019 World Bank report. These key informants were chosen purposively mainly due to the fact that an expert or a well knowledgeable informant can often enhance research significantly and more effectively than a randomly selected sample (Palys, 2003).

In the context of the purposive sample, the stakeholder sampling approach was specifically used to identify the key financial players who functioned as stakeholders and also made statements about the chapter's primary objective (Palys, 2008). As a result, their roles as governors and financial sector experts provided them with the necessary knowledge of Ghana's financial sector which justified their selection. Moreover, the researchers conducted a video content analysis by paying attention to videos that mentioned financial innovation in Ghana. The data were subsequently transcribed and categorised into themes, which present the foundational framework for most qualitative research (Corbin & Strauss, 2008). Further, before being grouped into major and subthemes, the responses were read several times. In the section which discusses financial inclusion for the marginalised, data presented were primarily narratives with contextual details quoted verbatim from the voices of participants (Mauthner & Doucet, 2003).

3.4 Ghanaian Context of Financial Innovation

Development in Ghana reveals that there has been a surge in financial technology in recent times. The country has seen an improvement in infrastructure and policy, driving several possibilities for financial institutions, and especially, the telecommunication industry to innovate, flourish and be sustainable in the financial space. Financial service products, including payments for goods and services, lending, insurance, remittances, and purchase of treasury bills (investments) currently benefit

[8] The YouTube videos chosen for this study were based on their focus and relevance to financial innovation in Ghana. The videos were chosen from face-to-face interviews on the YouTube channels of Consumer News and Business Channel (CNBC) Africa, the World Bank, and the International Center for Strategic Alliances (ICSA).

from the use of financial technology. The digitization of most payment services has brought with it some ease, convenience and security for individuals and businesses making instantaneous transactions. The sections which follow would attempt to bring these innovations to light from the perspective of the Central Bank and Ministry of Finance, and the traditional and emerging financial players such as the banks and the telecommunication industry.

3.4.1 *Product Development and Innovation—Interoperability and Strategic Roadmaps*

In 2018, the Central Bank, as part of its goal to encourage financial inclusion and a cash-lite economy, approved the applications of 108 banks and other financial institutions to introduce several payment products and services. Among the products and services that were permitted to be rolled out included agency banking services, issuance of payment cards (prepaid, credit and debit cards), cheque clearing services, electronic commerce, microloans and savings, and termination of inward remittances into bank accounts and mobile wallets (Bank of Ghana, 2018).

As part of its policy to transform its payment ecosystem, the Bank of Ghana implemented the first phase of the mobile money interoperability project in May 2018. The project facilitated the linkage of the mobile money services through the Ghana National Switch (gh-link) system to ease the transfer of funds across all the mobile money platforms. The three major national payment platforms; mobile money, bank accounts and E-zwich were interconnected in the second phase of the interoperability project in the latter part of the year. As part of the project, the three mobile money companies and participating banks were linked to the interoperability platform to provide services such as transfers across networks, wallet to bank transfers and E-zwich to wallet services (Bank of Ghana, 2018).

To provide a policy direction and guidelines to accelerate payments, the Central Bank launched the National Payment Systems Strategic Plan (NPSSP) (2019–2024), which is based on the improvements in digital technology in the country. The NPSSP was introduced to replace the 2014 Strategic Payment Systems Roadmap which had, among other goals, to move the economy to a cash-lite economy through the promotion and implementation of a more secured and efficient electronic payment system. The overall objective of the plan is to ensure efficient payments;

to advance financial innovation and to promote financial inclusion. It is particularly aimed at promoting electronic payments in the extensive informal sector where cash transactions constitute the main form of exchange.

The desire to introduce the NPSSP was anchored on the successes of the 2014 Strategic Roadmap. One positive outcome is the launch of the Ghana Interbank Payment and Settlement Systems (GhIPSS) Instant Pay (GIP) which allowed financial institutions to transfer to each other electronically. The provision of the gh-link card also allowed financial systems and institutions to interconnect with third-party financial service providers. Another significant result is the transformation of payment cards to the more enhanced and efficient Euro pay, MasterCard and Visa (EMV) chip and PIN standards. However, given the rate at which technology has improved globally and, in the country, the NPSSP was introduced to replace the 2014 Strategic Payment Systems Roadmap to take advantage of these well improved digital technologies. The plan seeks to reduce drastically, the paper-based payment forms to electronic payments to meet the ever-growing consumer demand for speed, convenience and accessibility of efficient payment systems (Bank of Ghana, 2019).

The Ministry of Finance is one of the key financial policy makers and implementers in the country. The outfit is responsible for introducing innovative ways to improve the financial system and to adopt strategies for broadening opportunities for financial inclusion. Over the course of the past four to five years, the government, through the Ministry of Finance has launched some policies to enhance the digital financial system in the country. Among these are the National Financial Inclusion and Development Strategy (NFIDS), the Digital Financial Services Policy and the Cash-Lite Roadmap.

The NFIDS, which was developed in collaboration with the World Bank, is aimed at increasing financial inclusion from the current 58% to 85% by the end of 2023 with the additional purpose of creating economic opportunities to reduce poverty. The Digital Financial Services Policy was also developed together with the Consultative Group to Assist the Poor (CGAP). Based on the foundation of the current fast developing technologies in the country, it seeks to improve the digital ecosystem which would contribute to the development of the private sector and the overall economy. The Cash-Lite Roadmap, also developed in collaboration with the Better Than Cash Alliance is targeted at developing a digital payments ecosystem that improves access to financial service, better regulation and supervision, and fostering consumer protection. In general, all three policies have one thing in common and seem to complement each other in the efforts to expedite the transition to a digital financial ecosystem; thereby, increasing the financial inclusion agenda of the country (Alorwuse, 2020).

3.4.2 *Financial Innovation by Banks and the Telecommunication Industry*

One of the ways banks have tried to reach their customers and also increase their customer base is increasing its branches. These branches are mostly networked. Networking bank branches help existing customers to access services at various places where their banks are located. Networking has been made possible by technology as it is vital for a bank branch to have the capacity to access the information of a customer no matter where the customer's main branch is located. Majority of the banks have also introduced internet banking services where customers can access most of the banking services online. According to the annual report, bank branches increased from 1,341 in 2016 to 1,491 in 2017 and then to 1,557 in 2018 (Bank of Ghana, 2018). The rise occurred in the period where the number of Universal banks reduced from 34 to 23 between 2017 and 2018. Hence, while the number of Universal banks was falling, the number of networked branches increased substantially.

The annual reports of some major universal banks in Ghana reveal several financial innovation services that have been introduced into the financial ecosystem. These financial improvements have contributed to broadening the financial space by including ordinary Ghanaians who may be unbanked, thus increasing financial inclusion. The following sections describe these specific innovations itemised in the annual reports of these universal banks.

NLA FastPay: This is an Agricultural Development Bank (ADB) initiative for the National Lottery Authority Lotto merchants and clients. It enables lotto merchants to easily purchase credits for their Terminal Portal Machines (TPM) and Portable Automation Data Process Machines (PADPM) point of sale terminals at any branch of the bank. Further, clients can redeem their winnings for all automation tickets. In this case, those who are unbanked but engage in lottery are automatically included in the financial ecosystem because of this innovative product (ADB, 2017).

ADB Bulk Payment: This is another initiative of ADB which provides a platform for customers with a secure, centralised electronic medium to initiate direct credit without necessarily visiting any bank branch (ADB, 2017).

ADB Visa Classic Prepaid Card: This is a Visa card that enables its owners to load and access the money instantly without necessarily having a bank account. Given this kind of product, customers can access bank products without necessarily opening a bank account with a physical bank (ADB, 2017).

International ATM Cards: An improvement in the ATM cards is the internationally recognised card which includes the visa and master cards. Most universal banks in Ghana have introduced these cards to enable customers to withdraw money from ATMs of other banks within and outside of Ghana. It also allows customers to shop online and pay for other services online. These cards also let customers pay for items in the card accepting supermarkets, restaurants and hotels, among others.

Bank Mobile App: Most banks in Ghana have developed tailor-made smartphone banking software. The mobile applications allow customers of banks to access bank statements, check their balance, airtime top-up, bill payment, pay for school fees, mobile money services, transfer funds, among others.

Mobile Short Codes: Banks such as GCB Bank and Guarantee Trust (GT) Bank have introduced mobile phone short codes to help its customers access some banking services. This service does not need a smartphone nor access to the internet making it convenient for customers to access some banking services no matter their location as long as they have network access.

Automated Teller Machines (ATMs): ATMs are computerised telecommunication devices which provide bank customers with access to financial transactions in public spaces. According to data from the International Monetary Fund (IMF) (2021), access to ATMs has increased over the years. A look at Fig. 3.1 shows that ATMs per 100,000 adults increased sharply between 2011 and 2013 and has further seen a steady increase from 2014 to 2018. The extent of ATM usage provides a clear signal of universal banks expanding their reach to their clients through technology. According to the Bank of Ghana (2018) figures, ATMs increased from 1,928 to 2,044 in 2017 and increased further to 2,139.

By far, the telecommunication industry has been one of the most successful initiators of financial innovation in the country. Their mobile

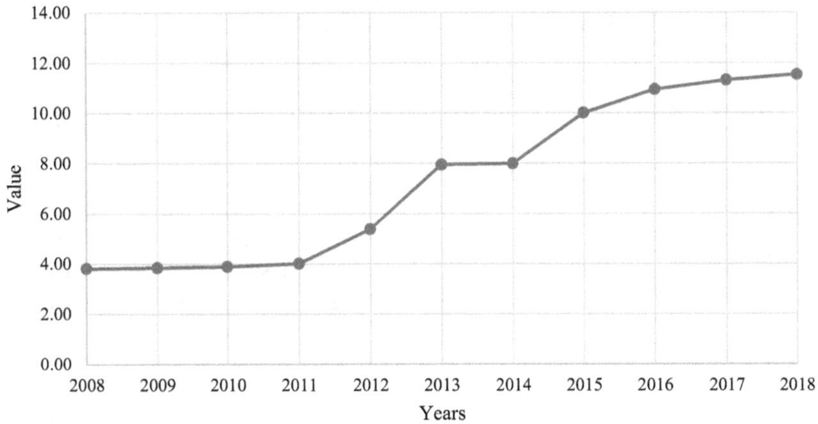

Fig. 3.1 Automated teller machines (ATMs) (per 100,000 adults) (*Source* International Monetary Fund [2021], Financial Access Survey)

money service has been very successful and has had considerable leverage in the financial industry. The mobile money service introduced in 2009 by the industry leader—MTN, has seen tremendous growth and the single largest contributor to the financial deepening and inclusion in the country (GMSA, 2020). In addition to payments of utilities, bills and cash transfers, the mobile money platform has also provided the means for financial and non-financial institutions to provide some additional services such as insurance and treasury bill purchases. According to data from the Bank of Ghana (2019), the total number of mobile voice service subscribers increased from 25.6 million to 40.01 million from 2012 to 2018. Within the same period, however, there was a corresponding increase in mobile money subscribers from just 3.8 million to over 32.6 million with active mobile money users increasing from a low of 345,434 to 13.1 million. The value of money transactions according to the central bank, therefore, increased substantially from GH¢594.12 million to GH¢223,207.23 million within the same period. The success has been recognised by global players like the World Bank, which has referred to Ghana as the fastest growing mobile money market in Africa.

3.5 Promoting Inclusion for the Financially Marginalised

In Ghana, financial inclusion appears to have provided disadvantaged people in rural places with the opportunity to acquire lower-cost and higher-quality financial services such as SMS banking, mobile money and others. During an interview with CNBC Africa, the Governor of the Bank of Ghana hinted that people living in rural areas are financially excluded from financial products and services and emphasised the use of digitalization as a major factor in narrowing the gap between the financially included and those currently excluded from the system. He phrased it this way:

> ...I had argued that digitalization is a game changer and for some of the big problems that we have in Africa, if you look at the financial sector, we have those that were financially excluded because of geography like the rural areas. We have large informal sectors that did not have access to finance etc...

To address this challenge, he went on to say that financial institutions like rural banks and microfinance enterprises are being developed in these areas to help fill the gap. He expresses himself clearly as follows:

> ...so now we try to address this issue by you know having the traditional banks opening up branches trying to set up other tiers of institutions such as micro finance companies and you know smaller rural banks in the villages and all that which is the way the traditional banking approach the issue of financial exclusion. (Ernest Addison, Governor, Bank of Ghana)

These remarks from the Governor of the Bank of Ghana suggests that financial inclusion is being by actively driven by institutions such as rural banks and microfinance institutions established in deprived communities.

Another crucial aspect of financial innovation, which is being used to extend financial inclusion is the interoperability mechanism established between mobile money platforms. This initiative has resulted in a number of advantages. For instance, individuals who were previously unable to access financial products and services are now able to do so due to the integration of the mobile money function into the financial system. This is how the Governor of the Bank of Ghana expressed it:

> Now with digital technology and mobile money in particular, we are able to reach everybody in the system more or less, and therefore, in a sense we're including all of these people in the financial sector... In other words, you can have transfers done between Mobile Money wallets and bank accounts. So, once you're able to reach to all those in the informal sector, who were not previously able to be part of the financial system, now, you can bring them all into the financial system through the linkage with Mobile Money accounts. (Ernest Addison, Governor, Bank of Ghana)

According to the response, the Bank of Ghana has expanded its reach with the support of the telecommunications industry to include persons living in economically disadvantaged areas. Furthermore, through the different mobile money platforms accessible in Ghana, financially excluded individuals have the opportunity to participate in the financial system. Consequently, the number of financially included people has increased significantly. This notable rise in financial participation is summarised as follows by the first deputy governor of the Bank of Ghana:

> In less than a decade, we have increased the number of financially included person from 41 percent of the total population in 2014 to 58 percent in 2017 and a significant proportion of the positive change emanated from mobile money patronage. For instance, the volume of mobile money transactions rose by more than 40 percent from 982 million participants to 1.4 billion transactions between 2017 and 2018.

Based on the aforementioned reaction, it is crucial to recognise that mobile money usage is a substantial contributor to the influx of financially disadvantaged groups into the financial system. According to a Senior Financial Sector Specialist and Co-author of the 2019 World Bank study, '*The financial sector in Ghana has grown rapidly, and with it the number of Ghanaians with access to formal financial services*'.

Aside government and central bank measures, financial technology businesses and insurance companies have also played a significant role in making financial services and products available to all Ghanaians. In this case, the First Deputy Governor of Bank of Ghana puts it this way: '*...Fintechs have also partnered with insurance companies to deliver digital insurance products to the underserved...*'.

In spite of the progress achieved so far, some actors have expressed concerns regarding the inclusion of women and people in underdeveloped regions in Ghana.

> But some groups have been left behind including women, rural areas and also some of the underdeveloped regions of Ghana. The government has facilitated access to formal financial services including mobile money digital switching solution, but more can be done to leverage financial inclusion using technology… (Carlos Vicente, Senior Financial Sector Specialist & Co-author of World Bank report)

He also stated that there is still opportunity for development, particularly in terms of technology. As a result, in the 2019 World Bank report, he made five recommendations on how the government can speed up the process of financial inclusion. During an interview with the World Bank, he issued the following statement:

> The report makes five recommendations on how government can accelerate progress towards creating financial inclusion. The first recommendation is the digitization of government and utility payments, and the second is the linking of informal groups to formal financial sector. The third is the promotion of expansion of access points including through agent banking. The fourth is building the financial capability of consumers and the final recommendation is the leveraging of data to improve access to financial services. (Carlos Vicente, Senior Financial Sector Specialist & Co-author of World Bank report)

Despite the diverse nature of the 2019 World Bank report's recommendations, a peculiar pattern emerges. These suggestions envision a digital framework in which everyone has access to an electronic financial system. By digitizing most business operations and governmental institutions, this should increase the number of people using financial products and services and, in the long run, turn the country into a '*paperless state*', as Vice President Mahamudu Bawumia proposed.

3.5.1 *Benefits of Financial Innovation*

In Ghana, the benefits of financial innovation are perhaps most obvious to the general public in terms of specific retail innovations such as debit and credit cards. These systems offer an efficient means of paying for goods and services, and obtain cash and account services, and provides significant benefit in terms of personal safety from targeted and highway robbery, and a safe and convenient way to make purchases by telephone and online respectively. This has aided in the reduction of banking inefficiencies (Frimpong, 2010) and the promotion of financial intermediation (Acquah, 2006). However, the benefits of innovation in the financial services sector extends well beyond innovative retail products to create a broader economy, create jobs, and improve the lives of Ghanaians. In the words of the First Deputy Governor, Bank of Ghana: '… *digital innovation is creating unprecedented opportunities for Africa to grow its economy, create jobs and transform people's lives…*'.

This signifies the confidence in new technologies as the solution to issues of underdevelopment, unemployment, and individual hardships within the sub region. The foregoing affirms the suggestion of a positive impact of technological innovations on the growth of African economies (Adeleye & Eboagu, 2019; Batuo, 2015). It also gives credence to Datta and Agarwal's (2004) claim that future growth in digital innovation is gradually being recognised as having a strong tendency to boost productivity and enhance economic growth.

Digitisation has also transformed the banking industry to improve access and deepen inclusion. It has introduced variety of products and services in the banking sector, according to the first deputy governor of the Bank of Ghana. He revealed that:

> … Indeed, digitization provides great potential for the banking sector to offer products and services that hitherto was not regarded as feasible…

The statement above implies a game-changing innovation in the banking sector such as e-banking, which enables them to transform how they connect with consumers (Cziesla, 2014).

In view of the projects and processes embarked upon by the Central Bank and the Ministry of Finance, it is evident that the government has been a major advocate in promoting financial innovation through digitalization. Particularly, the significant progress in the use of financial services is amplified by the First Deputy Governor of the Bank of Ghana, who stated that: '*…we are also witnesses to the tremendous impact of digitization of the financial sector. Ghana's progress in migrating to the electronic payments has been phenomenal*'.

It is important to note from the preceding remark that the transition from a paper-based to a technology-based system has paved the way for the financial sector to effectively utilise technology (Rose et al., 2009). Moreover, the Ghanaian government believes that leveraging technology to promote inclusivity can help to accelerate growth and alleviate poverty. In this instance, the Bank of Ghana's First Deputy Governor stated:

> The government of Ghana's commitment to formalize the Ghanaian economy through digitization is laudable, digitization is a key component of the 'Ghana Beyond Aid' agenda which is firmly anchored on leveraging technology to promote economic efficiency and inclusiveness for accelerated development and poverty reduction.

To ensure a safe and secure inclusion, the government has also launched many digital initiatives, such as a digital property addressing system and national biometric ID cards that will be linked to government agencies to facilitate identification and processing while also promoting confidence in business operations. The First Deputy Governor continues to emphasise that:

> ...Indeed, the president has shown keen interest and remains committed to building an inclusive modern society. This drive has seen the rollout of several digital initiatives such as the digital property addressing system, the national biometric ID cards which will be linked to the government agencies to help ease identification and processes and promote integrity and efficiency in business transactions. (Maxwell Opoku-Afari, First Deputy Governor, Bank of Ghana)

According to the above narrative, financial transactions are largely done virtually utilising e-banking services such as prepaid, credit and debit cards as a result of the financial sector's digitalization. As a result, it is critical to strengthen Ghana's identification and addressing system in order to track down malfeasance. This could enhance consumer confidence in the financial sector.

Currently, the pace of financial innovation has increased competition between traditional financial services and non-traditional firms such as financial technology companies (Fintechs) and mobile money providers. In a statement delivered at the digital banking meeting in January 2020, the Bank of Ghana's First Deputy Governor hinted that due to technology improvements, financial products and services have become more expansive. Consequently, there is more rivalry among banks and non-traditional providers such as the Fintechs.

> With the advent of technology, conservative banking operations has given way to dynamism and competition as innovative financial services and products are constantly being churned out. Currently, banks are not only competing among themselves for financial services but also with non-traditional providers such as financial technology companies, the fintechs...

In some cases, traditional and non-traditional financial providers collaborate to develop innovative techniques to make the financial industry more convenient and efficient. For example, as the First Deputy Governor of the Bank of Ghana put it, '*In the mobile money space, financial technology companies, the Fintechs, banks, mobile money operators have collaborated to introduce value-added services such as digital credit, savings and investment, online payment platforms, Unstructured Supplementary Service Data (USSD), mobile banking services, quick response codes, point of sales services, termination of inward remittances into mobile wallets or bank accounts, and agency banking services*'.

It is evident that banks and Fintechs are working together to expand Ghana's payment infrastructure. This indicates that individuals are more comfortable using electronic services such as ATM cards and various mobile money accounts than the traditional means of payment. The change has, however, made it possible for a greater reliance on technology (Rose et al., 2009). A rise in the ownership and use of mobile phones has seen individuals and businesses become more confident in adopting mobile-based digital payments services. Generally, mobile money has been seen as the most successful financial innovation in recent times.

Currently, financial innovation in Ghana has advanced beyond accepting payments and being merely payment and financial transactions-based to altering longstanding businesses such as restaurants, hotels, supermarkets and many more. Notably, the total volume of GhIPSS transactions recorded at the end of the first half of 2021 amounted to 70.2 million reflecting an increase of 142% over the volume noted for 2020.[9] This suggests that the Ghanaian financial system's payment structure is expanding. It is, therefore, evident that the proliferation of financial innovation has something to offer to diverse segments of the population. Such unique developments appear to be creating a thriving financial environment, along with fostering and strengthening financial services delivery.

[9] GhIPSS 2021 Half Year Performance Report: https://ghipss.net/phocadownload/newsletter/External-Newsletter-Vol-8.pdf.

3.5.2 Financial Innovation and Client Protection

Given the current pace of financial innovation, the need to sustain customers' interest in the financial sector is as urgent as the initial attempt to have them included. Financial services are becoming progressively riskier as a result of the vulnerability associated with digital and online technology, as fraud and other hazards endanger operations, financial soundness and stability, and public confidence in the financial system. Given the present rate of financial innovation, it is important to note that it has not been without its flaws. It is, therefore, worth noting that the implementation of financial innovation products such as mobile money has posed a number of obstacles such as financial innovation fraud and cyber threats.

According to the Director-General of the Cybercrime Unit of the Criminal Investigations Department (CID),[10] mobile money fraud is often committed by criminals who defraud innocent subscribers of various sums of monies, and 357 cases were reported in 2017[11] (Africa Press, 2019). The Minister of Communications[12] also added that almost half of Ghanaians who use any of the mobile money platforms have been targeted or are victims of mobile money fraud. Regrettably, only 10 per cent of the reported cases have been investigated and prosecuted (Africa Press, 2019). Such cases threaten the financial inclusion agenda of the government as highlighted by the words of the First Deputy Governor, '*the future of banking services is digitized and it brings opportunities as well as challenges*'. He continues to note that:

> There has also been an increased spate of complex cybersecurity threats which is posing risk to the cash-lite agenda...Despite the successes achieved so far, some consumers still lack awareness and confidence in electronic payments.

It is crucial to note that in the pursuit of attaining a cash-lite agenda, which has seen massive investments in the Ghanaian financial system's payment system infrastructure, the provision of a platform for digital

[10] Assistant Commissioner of Police (ACP) of the Ghana Police Service, Dr Gustav Herbert Yankson.

[11] The figure is reported to have declined in 2018.

[12] Ursula Owusu-Ekuful, Minister of Communications.

payments, such as mobile money interoperability has downsides such as being vulnerable to hackers.

To overcome these problems, some solutions have been proposed. The Bank of Ghana's First Deputy Governor offered a series of suggestions based on the aforementioned concerns. He puts it this way:

> The Bank of Ghana has therefore issued a cybersecurity directive for compliance by banks and payment service providers to build a robust and resilient digital ecosystem. To further accelerate the digitization process, stakeholders must work collectively to promote merchants' acceptance of digital financial products and services with particular emphasis on lower tier merchants such as small and micro enterprises.

Another option for improving the financial system and reducing the negative features of financial innovation is to increase trust and educate people. In this situation, the Bank of Ghana's First Deputy Governor stated that several financial service providers and stakeholders have been engaged in informing the public about recent developments in the financial sector. He worded it this way:

> The next key area is enhancing public confidence. In this direction, the various stakeholders are undertaking continued consumer education and financial literacy to build confidence and help in fraud reduction. GhIPSS and stakeholders in the payments industry are also building a fund to support consumer education. Also, on a regular basis, payment service providers are educating the public on how to avoid scams in the use of newly deployed payment products and services.

The Ghana Interbank Payment and Settlement Systems (GhIPSS), which the central bank of Ghana established to manage interoperable payment system infrastructures for banks and non-bank financial entities, has made payment transactions easier, as stated above. This could be linked to the investment in raising awareness about new payment options in order to avoid fraudulent activities. As a result, the degree of consumer confidence rises.

Furthermore, it is critical to remember that the financial sector's transition to digitisation and innovation must satisfy consumer demand. This should, in the long run, increase financial sector efficiency and expand chances for inclusiveness in financial products and services, thereby promoting financial inclusion across the country. Similarly, the Bank of Ghana's First Deputy Governor noted briefly:

> However, the transition to a digital economy requires reorientation of the sector to meet consumer demand...we can harness digitization to promote transactional efficiency in the delivery of financial services as well as scale-up and broaden financial assets to all segments of the society. Together, we must liberate on these opportunities to deliver Financial Inclusion across our various countries and on the continent at large.

According to the above statement, the financial sector should be revamped to satisfy consumer demand and to expand the financial system to accommodate marginalised individuals. This could assist to mitigate some of the issues that arise as a result of the introduction of new technology in Ghana's financial sector.

3.6 Conclusion

This chapter presented insights into the nature and extent of financial innovation in Ghana. It highlights how Ghana is adopting, revising and implementing some of the identified financial innovation strategies to give opportunities for a greater number of people to be financially included without the traditional means of opening or having a bank account. The information gathered suggest that the central bank, financial institutions and mobile telecommunication operators are all contributing to the financial innovation drive in Ghana. While successes have been achieved in the use of ATMs, domestic and internationally recognised ATM cards, mobile short codes and bank-specific mobile application software, the most considerable momentum is recorded in mobile money subscription and transactions. Overall, mobile money technology has contributed significantly to reducing the number of the financially excluded.

Around the world, various financial innovative projects are also being pursued or targeted for future implementation. As these projects and initiatives are rolled out, they present a challenge, particularly for developing economies, on what could be attained in the future to make the financial system more efficient, reliable and inclusive. The immense drive, particularly in the Asian sub-continent, provides learning opportunities on how domestic and cross-border payments could be achieved through partnerships.

In the Ghanaian context, the launch of the mobile phone interoperability project has greatly empowered most small businesses in the way they conduct, develop and integrate their businesses with other supporting partners in terms of payments to creditors or utility providers. The effort has resulted in a relatively efficient system compared to the time-consuming paper-based and human-faced interactions of the earlier regime, which excluded a larger percentage of people in the informal sector. While outlining the process and product-enhancing features of financial innovation, this chapter highlights the critical issue of the threat posed by financial innovation fraud and its implications for client protection. These challenges call for robust external and internal risk assessment mechanisms to combat such threats.

For financial innovation to succeed in promoting financial inclusion, which should enhance economic growth and prosperity in Ghana, it would require more robust and aggressive forms of protecting clients' data and deposits. Active client protection measures should improve and maintain financial confidence in the fast-paced, innovative systems geared at making the use of financial services less cumbersome and intractable and more rewarding for clients from all aspects of life.

References

Abor, J. (2005). Technological innovations and banking in Ghana: An evaluation of customers' perceptions. *IFE Psychologia: An International Journal, 13*(1), 170–187.

Acquah, P. (2006). The emerging Ghanaian banking environment. *BIS Review, 75*, 1–2. Retrieved from https://www.bis.org/review/r060825e.pdf

Adeleye, N., & Eboagu, C. (2019). Evaluation of ICT development and economic growth in Africa. *NETNOMICS: Economic Research and Electronic Networking, 20*(1), 31–53.

Africa Press. (2019). *Momo fraud: How scammers steal money*. Retrieved from https://www.africa-press.net/ghana/all-news/momo-fraud-how-scammers-steal-your-money

Agricultural Development Bank (ADB). (2017). *2016 annual report and financial statements*. ADB, Accra. Retrieved from https://www.agricbank.com/wp-content/uploads/2018/04/2016-ADB-Annual-Report.pdf

Alipay. (2015). *A powerful partnership: Alipay digital payment with cybersource*. Retrieved from https://global2.cybersource.com/content/dam/cybersource/en-EMEA/documents/Alipay_Partner_DS.pdf.

Alorwuse, W. (2020). *Ghana introduces new digital financial inclusion policies: A catalyst for growth.* Strategic Impact Advisors. Accessed from https://www.siaedge.com/news/2020/6/23/ghana-introduces-new-digital-financial-inclusion-policies-a-catalyst-for-growth

Avlonitis, G. J., Papastathopoulou, P. G., & Gounaris, S. P. (2001). An empirically-based typology of product innovativeness for new financial services: Success and failure scenarios. *Journal of Product Innovation Management, 18*(5), 324–342.

Bank of Ghana. (2018). *Annual Report 2018.* Bank of Ghana, Accra. Retrieved from https://www.bog.gov.gh/wp-content/uploads/2019/07/AnnRep-2018.pdf

Bank of Ghana. (2019). *Payment System Statistics 1st Quarter 2019.* Accra, Bank of Ghana. Retrieved from https://www.bog.gov.gh/wp-content/uploads/2019/08/PAYMENT-SYSTEM-STATISTICS-Ist-Quarter-2019.-1.pdf

Bank of Lithuania. (2019). *CENTROlink report.* Retrieved from https://www.lb.lt/en/news/payment-system-centrolink-marks-instant-payments-era-in-lithuania

Bank of Portugal. (2018). *Regulators and Portugal Fintech Association create a platform to support technological innovation in the financial sector* [Press Release]. Retrieved from https://www.bportugal.pt/en/page/fintech

Bank of Thailand. (2019). *Bank of Thailand Inthanon phase 1 report.* Thailand. Retrieved from https://www.crowdfundinsider.com/2019/05/147194-project-inthanon-bank-of-thailand-%09pushes-forward-with-central-bank-digital-currency/

Batuo, M. E. (2015). The role of telecommunications infrastructure in the regional economic growth of Africa. *The Journal of Developing Areas*, 313–330.

Chapman J., Garratt R., Hendry S., McCormack A., & McMahon W. (2017). *Project Jasper: Are distributed wholesale payment systems feasible yet?* Bank of Canada, Financial System Review.

Child, J., Faulkner, D., Tallman, S., & Hsieh, L. (2019). *Cooperative strategy: Managing alliances and networks.* Oxford University Press.

Corbin, J., & Strauss, A. (2008). *Basics of qualitative research: Techniques and procedures for developing grounded theory.* Sage Publications.

Cziesla, T. (2014). A literature review on digital transformation in the financial service industry. In *BLED 2014 Proceedings.* 18. Retrieved from https://aisel.aisnet.org/bled2014/18

Danske Bank Group. (2007). *Danske Bank Annual report 2007.* Retrieved from https://www.annualreports.com/HostedData/AnnualReportArchive/d/danske-bank_2007.pdf

Datta, A., & Agarwal, S. (2004). Telecommunications and economic growth: A panel data approach. *Applied Economics, 36*(15), 1649–1654.

Deutsche Bank. (2014). Fintech-The digital (r)evolution in the financial sector. *Deutsche Bank Research*. Retrieved from https://www.dbresearch.com/PROD/RPS_EN-PROD/PROD0000000000451941/Fintech_%E2%80%93_The_digital_(r)evolution_in_the_financia.pdf?undefined&realload=ACX/n~3/TwqNu81YAHDrqOw6QK8~CSOIhH~jaNIJGdGBu/mwpKRMtrr2p~dIkONg

Frimpong, J. M. (2010). Investigating efficiency of Ghana banks: A non-parametric approach. *American Journal of Scientific Research, 7*, 64–76.

GMSA. (2020). *MTN MoMo pay merchant payments: Expanding women's mobile money use in Ghana*. Retrieved from https://www.gsma.com/mobilefordevelopment/wp-content/uploads/2020/05/MTN-MoMo-Pay-Merchant-Payments-Expanding-Female-Mobile-Money-Usage-in-Ghana.pdf

Hong Kong Monetary Authority. (2018). *The launch of faster payment systems*. Hong Kong Monetary Authority [Press releases]. Retrieved from https://www.info.gov.hk/gia/general/201809/17/P2018091700743.htm

International Monetary Fund. (2021). *Financial Access Survey*. IMF. Retrieved from https://data.imf.org/?sk=E5DCAB7E-A5CA-4892-A6EA-598B5463A34C&sId=1460043522778

Mauthner, N. S., & Doucet, A. (2003). Reflexive accounts and accounts of reflexivity in qualitative data analysis. *Sociology, 37*(3), 413–431.

Mention, A. L., & Torkkeli, M. (2012). Drivers, processes and consequences of financial innovation: A research agenda. *International Journal of Entrepreneurship and Innovation Management, 16*(1/2), 5–29.

Nkem, I. S., & Akujinma, A. F. (2017). Financial innovation and efficiency on the banking sub-sector: The case of deposit money banks and selected instruments of electronic banking. *Asian Journal of Economics, Business and Accounting, 2*(1), 1–12.

Palys, T. S. (2003). *Research decisions: Quantitative and qualitative perspectives*. Thomson Nelson.

Palys, T. (2008). Purposive sampling. *The Sage Encyclopedia of Qualitative Research Methods, 2*(1), 697–698.

Pandy, S., & Crowe, M. (2019). *Innovation across mobile payments: Alternative checkout models, Chinese wallets, Fintechs and mobile operators*. Federal Reserve Bank of Boston. Retrieved from https://www.bostonfed.org/publications/mobile-payments-industryworkgroup/innovation-across-mobile-payments.aspx .

PricewaterCooper. (2016). *Financial service technology 2020 and beyond: Embracing disruption*. PwC. Retrieved from https://www.pwc.com/gx/en/financial-services/assets/pdf/technology2020and-beyond.pdf

Rose, P. S., Marquis, M. H., & Lu, J. (2009). *Money and capital markets*. China Machine Press.

Schindler, J. W. (2017). *FinTech and financial innovation: Drivers and depth* (FEDS Working Paper No. 2017-81). Retrieved from https://papers.ssrn.com/sol3/papers.cfm?abstract_id=3029731

Seven Bank. (2016). *First in Japan! LINE pay deposits and payments to become available at ATMs.* Seven Bank Ltd – Japan. Retrieved from www.sevenbank.co.jp/english/ir/pdf/2016/2016111801_E1.pdf

South African Reserve Bank. (2018). *Project Khokha: Exploring the use of ledger technology for interbank payment settlements in South Africa.* Project Khokha Fintech Report. Retrieved from https://www.resbank.co.za/Lists/News%20and%20Publications/Attachments/8491/SARB_ProjectKhokha%2020180605.pdf

Tahir, S. H., Shah, S., Arif, F., Ahmad, G., Aziz, Q., & Ullah, M. R. (2018). Does financial innovation improve performance? An analysis of process innovation used in Pakistan. *Journal of Innovation Economics Management, 3*, 195–214.

Vodafone. (2019). *M-Pesa.* Vodafone UK. Retrieved from https://www.vodafone.com/what-we-do/services/m-pesa.

CHAPTER 4

Financial Inclusion and Monetary Policy Effectiveness in Ghana

William Godfred Cantah, James Atta Peprah, and Paul Owusu Takyi

4.1 Introduction

One key objective of the Ghanaian economy over the last three decades has been to move from a lower-middle income economy to a higher-income economy in the shortest possible time. This objective can only be achieved in a stable macroeconomic environment which allows businesses to thrive. One of the key indicators of macroeconomic stability is the rate of inflation. The attainment of lower rates of inflation is an indication that general prices in the economy rise at a slow pace making it easy for

W. G. Cantah (✉)
Department of Data Science & Economic Policy, School of Economics, University of Cape Coast, Cape Coast, Ghana
e mail: william.cantah@ucc.edu.gh

J. A. Peprah
Department of Applied Economics, School of Economics, University of Cape Coast, Cape Coast, Ghana
e-mail: jpeprah@ucc.edu.gh

J. A. Peprah et al. (eds.), *Financial Sector Development in Ghana*, Palgrave Macmillan Studies in Banking and Financial Institutions,
https://doi.org/10.1007/978-3-031-09345-6_4

businesses to effectively plan their investments which would eventually propel the growth of the economy. In Ghana monetary policy has been used to maintain price stability and steer the economy towards a path of stable growth.

The potency of monetary policy to influence price stability and propel growth of the economy largely depends on the proportion of populace who are included (or who have accounts) in the formal financial system. This is because monetary policy would have limited or no impact on the intertemporal consumption decisions of people who are financially excluded (Anarfo et al., 2019). Thus, well-developed non-excluding financial system is very critical for implementing monetary policy effectively to achieve the developmental aspirations and goals of the Ghanaian economy. Financial inclusion is seen as a critical feature of financial development (Le et al., 2019), hence a well-developed financial system can be seen as one that includes majority of its citizen especially the poor, agricultural households as well as micro-entrepreneurs (Peprah et al., 2021). This implies that financial inclusion is very critical for the effective transmission of monetary policy. This chapter examines the impact of key component of financial sector development (financial inclusion) on the effectiveness of monetary policy in Ghana. This is particularly important due to the fact that an effective implementation of monetary policy would provide a strong foundation for the growth and development of the country, which would also go a long way to enhance further development of the financial sector (see Anarfo et al., 2019). This chapter brings to bear one of the key benefits of financial sector development and how it could enhance the stability of the macro economy to propel national development. The role of financial inclusion in the transmission of monetary policy has occupied the attention of policymakers, scholars, and researchers in the last decade. This is because an increase (or reduction) in the level of financial inclusion could have implications for monetary policy effectiveness and macroeconomic stability as whole. This study thus, examines how financial inclusion affects the transmission of monetary policy in Ghana.

P. O. Takyi
Department of Economics, Kwame Nkrumah University of Science and Technology, Kumasi, Ghana
e-mail: paul.takyi@knust.edu.gh

Related studies have largely focused on the link between monetary policy and financial inclusion (see for Anarfo et al., 2019; Evans, 2016; Lapukeni, 2015; Lenka & Bairwa, 2016; Mbutor & Uba, 2013). These studies were not without some challenges: first, some of the studies were panel studies focusing on different countries with different monetary policy frameworks. That is these panel studies tend to mix economies that practice Inflation Targeting (IT) framework and Monetary Targeting (MT) framework in the same model. This makes it relatively difficult to isolate the effective role of financial inclusion in monetary policy effectiveness for the different policy frameworks. Whilst increase in financial inclusion may be very crucial in the transmission of monetary policy in IT economies (due to its focus on interest rates as the main tool of monetary policy) (see Samba, 2018), financial inclusion may not have the same effect on monetary policy transmission in MT framework.

Second, a number of these studies used very weak variables to represent for monetary policy and financial inclusion and example of such papers include (Evans, 2016; Lapukeni, 2015; Mbutor & Uba, 2013). However, inflation is only an outcome variable and not a policy variable (see Anarfo et al., 2019). Measuring monetary policy effectiveness should thus, focus on the ability of monetary policy to adequately affect movements in inflation rate rather using movements in inflation as the measure of monetary policy. In addition to the challenge of not using appropriate proxy for monetary policy, a number of these studies usually proxy financial inclusion by just a single variable. Financial inclusion has been identified to have two dimensions, usage and access. Hence any measure of financial inclusion that ignores one of these dimensions may not clearly be measuring the true level of financial inclusion. Fourth, these studies failed to identify the extent to which changes in the level of financial inclusion affects monetary policy to cause a change in the level of price stability in an economy. Finally, they have mostly relied on the use of either VAR or Panel VAR to investigate the link between financial inclusion and monetary policy, however VAR is known to be exposed to the Lucas critique.

To address these gaps, we examine the role of financial inclusion and monetary policy from the perspective of an inflation targeting economy (i.e., Ghana). These gaps are addressed by first, focusing on an inflation targeting economy, hence it presents a unique situation without any form of contamination from a monetary targeting economy. Second, we appropriate proxies for monetary policy and financial inclusion employing the Monetary Policy Rate (main tool of monetary policy in Ghana) as a proxy for monetary policy and financial inclusion index using Principal Component Analysis (PCA) Third we measure monetary policy effectiveness by analysing the effect of the interaction between financial inclusion and monetary policy on inflation. We conduct counterfactual analysis to identify the role of financial inclusion in the transmission of monetary policy. The present study also employs a Structural VAR model (which as an advantage of addressing the Lucas critique) to empirically analyse the effect of financial inclusion on monetary policy through the use of an interaction term (interaction of financial inclusion and monetary policy) and also doing counterfactual analysis on the role of financial inclusion for Ghana. We argue that higher levels of financial inclusion would work to increase the elasticity of spending to interest rate and hence increase the effectiveness of monetary policy.

The rest of the chapter is organised as follows: the second section examines related literature on financial inclusion and monetary policy. This is followed by the data and methodology for the research. The fourth section focus on analysis and discussions of the empirical results and the last section presents the conclusions and policy implications of our results.

4.2 Review of Related Literature

De Bondt (2000) argued that monetary policy works mainly through two main channels: direct monetary transmission and interest channels. The direct monetary transmission involves the manipulation of the supply of money in the economy to affect spending levels within the economy and ultimately affect general price levels. With the interest rate channel, the central bank makes adjustments to the policy rate which tends to affect lending rate in the money market which then affects cost of borrowing (affecting investment and consumption decisions in the economy). This would then affect aggregate demand and supply and eventually affect price level. Implicit in the use of the interest rate channel is the assumption that all economic agents in the economy have access to financial services. This assumption is inherently erroneous in a number of developing economies where a number of agents lack access to financial services and the few that have access to financial services are mostly unable to access them. Thus, the success of the interest rate channel largely depends on the extent of financial inclusion in an economy. Brownbridge et al. (2017) argued that high levels of financial exclusion tend to reduce the elasticity of interest rate of private spending which also dampens the effect of the interest rate on price stability.

The IT policy framework (adopted by Ghana and South Africa) is based on the Taylor's rule of monetary policy which requires monetary policy makers to raise nominal interest rate by percentage that exceed an increase in the inflation rate (Woodford, 2001). Brownbridge et al. (2017), however argued that, the expected effect of the Taylor's rule could be weakened, if private sector spending and for that matter aggregate demand is interest inelastic. The effectiveness of the Taylor's rule under an inelastic interest regime would require Central Banks to apply more aggressive changes to the policy rate. This implies that the exclusion of a large proportion of the adult population from the formal financial sector could have implications for the implementation of monetary policy via the interest rate channel, especially for IT economies which interest rate is the main instrument of monetary policy. The need to understand the implication of financial inclusion for monetary policy implementation has culminated into a number of empirical activities in the area.

Mbutor and Uba (2013) examined the relationship between financial inclusion and monetary policy for the Nigerian economy for the period 1980–2012 using a VAR model. Their key finding indicated that a percentage increase in the ratio of total loans and advances reduced inflation by 0.01%. They therefore concluded that increase in the level of financial inclusion supports the achievement of monetary policy objective. The use of inflation in their work is quite misleading. Inflation in itself is an outcome variable and not the policy variable, analysing the impact of financial inclusion on monetary policy requires an examination of how financial inclusion affects attainment of the objectives of monetary policy. Brownbridge et al. (2017) maintains that the model used by Mbutor and Uba (2013) lacked theoretical backing.

Evans (2016) analysed the impact of financial inclusion on monetary policy effectiveness for 15 Sub-Saharan African countries (including Ghana and South Africa) using a Vector Error Correction Model (VECM). Like Mbutor and Uba (2013), he also used inflation (an outcome variable) as a measure of monetary policy effectiveness. The study concluded that financial inclusion was not a key driver of monetary policy effectiveness in Africa, but rather monetary policy effectiveness is what drives financial inclusion. Apart from not using the right variable to measure the effectiveness of monetary policy, the study combined countries with different monetary policy regimes which used different monetary policy instruments. The effect of financial inclusion on monetary policy would vary depending on the policy instrument due to the differences in the transmission channels of these instruments. For example, economies that target money supply (with money supply as the main policy instrument) would have transmission channels which may differ from IT economies. Thus, combining countries with different monetary policy framework could only lead to erroneous conclusions.

Lenka and Bairwa (2016) employed fixed effect model to examine the effect of financial inclusion on monetary policy in the South Asian Association for Regional Cooperation (SAARC). Lenka and Bairwa (2016) just like the previous studies used inflation as their measure of monetary policy. An interesting aspect of their study was the use of an index of financial inclusion. Their study found a negative relationship between financial inclusion and inflation.

Brownbridge et al. (2017) examined the impact of financial inclusion on the interest rate channel of monetary policy transmission mechanism. The study focused on Ghana, Mauritius, South Africa and Uganda using a Panel VAR model. The study assessed the impact of financial inclusion on monetary policy by classifying the economies into high inclusion countries (Mauritius and South Africa) and low inclusion countries (Ghana and Uganda) and analysed the impact of interest rate shock on consumer price index. The conclusion was that the response of (consumer price index) CPI to changes in interest rate was stronger in countries with high levels of inclusion than countries with lower rates of financial inclusion.

The study period for this study (200Q1–2016Q4) meant that the study combined different monetary policy regimes in their analysis. South Africa adopted IT in 2000, Ghana 2002, Mauritius 2006 and Uganda 2011. Thus, the sample period of the study combined different policy regimes of these countries which makes policy attribution a bit difficult. In addition, the study did not control for financial inclusion in their model (it only separated the countries into high and low inclusion economies), it become difficult for the study to attribute the strong impact of interest rate on CPI to the level of financial inclusion since other economic condition could have also played a key role. Furthermore, the study did not do any form of counterfactual analysis to identify the actual impact of financial inclusion on monetary policy effectiveness. In addition, the study was just like most of the existing studies that examined financial inclusion and monetary policy effectiveness that used the VAR model which exposes it to the Lucas critique.

In a more recent study by Anarfo et al. (2019) the authors examined the bi-causal relationship between financial inclusion and monetary policy for sub-Saharan African countries using a Panel VAR for the period 1990–2014. The study used interest rate as a proxy for monetary policy and found a bi-causal relationship between monetary policy and financial inclusion but fails to clearly identify how financial inclusion affects monetary policy effectiveness. This is because interest rate is the instrument of monetary policy and not the object, understanding the effectiveness of monetary policy requires an analysis of the how financial inclusion affects the transmission of monetary policy to inflation and output.

The review of the empirical literature has revealed that number of existing studies have failed to examine the extent to financial inclusion affects the transmission of monetary policy. in most cases these studies tend to either focus on the effect of financial inclusion on inflation or the effect of inclusion on monetary policy itself without examining how inclusion affects monetary policy to affect inflation, the main object of monetary policy. The present study, therefore, departs from the existing studies by examining how financial inclusion affects the transmission of monetary policy to affect inflation. To achieve this, the study employs a structural VAR model which is free of the Lucas critique to analyse the impact of financial inclusion on monetary policy effectiveness by interacting financial inclusion with the monetary policy instrument to examine its impact on inflation and output. The study goes a step further to do counterfactual analysis to understand the specific impact of financial inclusion on the effectiveness of monetary policy.

4.3 Methodology and Data

4.3.1 Theoretical Model

Given the fact that the implementation of IT policy framework is based on the Neo-Keynesian model, the theoretical framework of the study was derived from the aggregate demand theory which is represented by the IS curve, aggregate supply curve captured by the Philips curve and the uncovered interest rate parity (UIP) equation. The study extends these equations by including an equation for average lending rate in the financial market and the level of financial inclusion. The model is closed by describing the monetary policy rule which is based on the Taylor's inrule.

The theoretical framework was derived from the work of Hove et al. (2015).

The nominal exchange rate is developed based on the UIP relation which relates the variations in the exchange rate to domestic and foreign exchange rate differentials:

$$s_t = s_{t-1} + \beta_1 (r_t - r_t^*) + \beta_2 d_t + \varepsilon_{st} \tag{4.1}$$

where, β is a parameter to be estimated, r_t is the domestic interest rate and r_t^* is the interest rate of foreign country and $\lceil_t$ is the budget deficit. ε_t is the exchange rate shock.

The aggregate demand curve is represented by the output gap equation which is represented by:

$$y_t = \phi_0 + \phi_1 y_{t-1} + \phi_2 \int_t + \phi_3 \lceil_t + \varepsilon_{yt} \tag{4.2}$$

where y_t is the output gap and ϕs are parameters and ε_{yt} is the disturbance term.

Aggregate supply curve is represented by the Philips curve:

$$\pi_t = \alpha_1 \pi_{t-1} + \alpha_2 y_t + \alpha_3 \mathcal{F} D_t + \alpha_4 \int + \varepsilon_{\pi t} \tag{4.3}$$

The monetary policy equation is based on the Taylor rule in which policy makers adjust interest rate in response to changes in in the rate of inflation, the output gap, and the exchange rate.

$$r_t^p = \lambda_1 r_{t-1}^p + \lambda_2 y_t + \lambda_2 \pi_t + \lambda_1 \int_t + \varepsilon_{rp_t} \tag{4.4}$$

where r_t^p captures the monetary policy rate and ε_{rp_t} is the disturbance or the error term. The specification in Eq. (4.4) implies that the central bank reacts to changes in inflation, exchange rate and output gap while keeping some degree of interest rate smoothing. The study also introduces an equation that captures response of the money market changes with Eq. (4.5)

$$r_t = \omega_1 r_{t-1} + \omega_2 y_t + \omega_3 \pi_t + \omega_4 \int_t + \omega_5 r_t^p + \varepsilon_{rt} \tag{4.5}$$

where r_t is the market interest rate, which respond to changes in the output gap, inflation rate, exchange rate and changes in the monetary policy rate, ε_{rt} is the disturbance term. In addition, the present study argues that the decision to be financially included in the financial sector would be influenced by stability of macroeconomy and the interest rate in the economy. Thus, the financial inclusion at the macro level would be determined by the following equation:

$$f = \theta_1 f_{t-1} + \theta_2 r_t + \theta_3 y_t + \theta_4 \pi_t + \theta_5 \int_t + \theta_6 r_t^p + \varepsilon_{ft} \tag{4.6}$$

where f is financial inclusion and θs are parameters to be estimated. ε_{ft} is the disturbance term. The study would also include an interacted term (interaction between monetary policy and financial inclusion). The dynamic responses of all variables are analysed under the inflation targeting regime where the central bank alters the policy rate to keep the inflation rate on target.

Equations (4.1) to (4.6) are summarised in the following matrix equation

$$\mathcal{A}\mathcal{Y}_t = \mathcal{X} + \mathcal{B}\mathcal{Y}_{t-1} + \epsilon_t \tag{4.7}$$

where $\mathcal{Y}_t$ is a vector of endogenous variables, which include fiscal deficit, exchange rate, output gap, inflation rate, monetary policy rate, lending rate and financial inclusion, $\mathcal{A}$ and $\mathcal{B}$ are vector & matrix of coefficients respectively. The interaction between monetary policy and financial inclusion is also added to the model.

Due to the correlation between the variables and the error term, the structural Eq. (4.1) cannot be estimated directly. The information in the system can be retrieved by estimating the reduced form equations which can actually be estimated. This can be done by pre-multiplying Eq. (4.1) by $\mathcal{A}^{-1}$ to arrive at the reduced form equation below:

$$\mathcal{Y}_t = \mathcal{F} + \Gamma\mathcal{Y}_{t-1} + \mu_t \tag{4.8}$$

where $\mathcal{F} = \mathcal{A}^{-1}\mathcal{X}, \Gamma = \mathcal{A}^{-1}\mathcal{B}$, and $\mu_t = \mathcal{A}^{-1}\epsilon_t$ is the reduce form residual vector, which is assumed to be white noise.

4.3.2 *Empirical Model and Description of Variables*

The paper considers the reduced form VAR model from Eq. (4.8). However, in order to improve the econometric properties of the model for the purposes of estimation, the study follows Hove et al. (2015):

$$\mathcal{Y}_t = \mathcal{F} + \Gamma(L)\mathcal{Y}_t + \mu_t \tag{4.9}$$

where, $\mathcal{Y}_t = FD_t, EX_t, CIEA_t, INF_t, MPR_t, LEN_t, FIN_t MF_t$, is a vector of all variables. The variables employed in the model were all derived from the theoretical model. Hence, all variables included in the empirical model based on theory.

FD_t represents fiscal deficit, which is measured by the ratio of GDP it was included in the model to capture the effect of fiscal policy on price stability and economic activities, EX_t is the nominal exchange rate measured as exchange rate between the domestic currency and the US dollar, $CIEA_t$ is the monthly growth of economic activities. The output was proxied by industrial production index for both countries. INF_t is the inflation rate year on year percentage changes in the consumer price index. MPR_t is monetary policy rate as announced by the central banks of the respective countries. LEN_t is the average lending rate by commercial banks. FIN_t is financial inclusion and MF_t is the interaction between monetary policy and financial inclusion. The study employed the principal component analysis to generate an index for financial inclusion. The variables used for the computation of the index are savings and time deposits, number of ATM per 100, 000 adults, outstanding loans by commercial banks as a percentage of GDP and commercial bank branches. Details of the PCA results are provided in the appendix. The study also interacted monetary policy with financial inclusion to identify the extent to which financial inclusion affects the transmission of monetary policy.

4.3.3 *Structural Vector Autoregressive Model (SVAR)*

The study used an 8-variable SVAR model to quantify the response of inflation and output gap to the interaction between monetary policy and financial inclusion in Ghana. The SVAR model was chosen due to the fact that it allows the use of a non-recursive structures based on relevant economic theory and empirical literature to impose restrictions on the parameters of matrices A and B (Bernanke, 1986; Cushman & Zha, 1997; Hove et al., 2015; Sims, 1986). In our framework, the variance–covariance matrix of the disturbances derived from $\mu_t = A^{-1}\epsilon_t$ is used to identify the structural shocks in the system. The vector of structural shocks is orthogonal (they are serially uncorrelated) whereas matrix A is allowed to have off-diagonal elements. Perhaps, this procedure is more appropriate to use when attempting to discriminate among structural hypotheses.

The identification restriction is based on the assumption that expenditure decisions by the central government in the short run is exogenous. As such, changes in fiscal deficit have the potential to affect domestic output gap, exchange rate, inflation rate monetary policy rate and interest rate., However all these variables are not expected to affect fiscal deficit in the short run. Changes in output due to increased aggregate demand could affect exchange rate (since most African economies tend to be import dependent) and the rate of inflation supply may not immediately change to meet the changes in aggregate demand. Thus, changes in the output gap could affect the exchange rate, inflation rate and the policy rate in the short run. The effect of depreciation or appreciation of the domestic currency is usually transmitted into the economy via the inflation channel. The changes in the rate of inflation call for response from monetary policy makers who manipulate the policy rate and the effect of the changes in the policy rate is transmitted back to the economy through its impact on domestic lending rate. The domestic lending rate, we argue that, would influence the decision of individuals in the economy to join the financial sector.

4.4 Data

The data used for the study spans from 2002 to 2018 and was sourced from secondary sources. The choice of 2002 was mainly because it marked the start of the informal implementation of the inflation targeting framework. Hence using 2002 as the start period would eliminate the possible effect of monetary targeting which was practiced before the IT framework was adopted. We obtained data from the World Development Indicators of the World Bank and the websites of the Central Bank of Ghana. Monthly data on inflation, exchange rate, lending rate, monetary policy rate, savings & time deposits and composite index of economic activities (CIEA) were obtained from the website of Bank of Ghana. Data on fiscal deficit was sourced from the Ministry of Finance, Ghana. The study used four main variables (which capture accessibility and usage) to measure financial inclusion. Accessibility was proxied by the number of commercial bank branches per $1000km^2$ (CBB) and the number of ATMs per $1000km^2$ (ATM). Usage was proxied by outstanding loans by commercial banks (OLCB) and savings and time deposits. All these variables were sourced from the Bank of Ghana and the World Development Indicators in annual series. However due to the fact that the rest of the data set employed were all in monthly series, the study used the method proposed by Denton (1971), for the interpolation of time series from a lower frequency to a higher frequency without affecting the behaviour of the series. The Denton (1971) approach finds an interpolated series by relating a higher-frequency indicator to the lower-frequency benchmark series. The Denton interpolation preserves the movement and behaviour of the original series. The Denton approach employed in this study ensures that the average of the interpolated monthly series give a figure that is equal to the annual series.

4.4.1 *Principal Component Analysis (PCA)*

Since the measures of financial inclusion has different dimensions, the study used PCA to reduce the dimensions of the data to generate a single variable for financial inclusion. Jolliffe et al. (2003) maintained that the most prominent variation of the original variable is captured in the first few principal components. The PCA, thus becomes a useful tool for identifying the structure of relationship among variables or reducing the number of variables included. The attributes of the variable been measured (in this financial inclusion) is summarized by a number of mutually exclusive independent principal components with each being the weighted average of the underlying variables. The main advantage involved in the use of the PCA approach is the fact that the weights of the index are based on the inner correlation of all individual measures.

The results of the PCA are shown in Table 4.3 of the Appendix. The eigenvalues of the first and second principal components were the only two that was greater than 1. However, the first component was found explain a little over 70% of the total variance whilst the second component explained just about 26% of the total variance. The Kaiser–Meyer–Olkin (KMO) measure of sampling adequacy which was used to analyse the appropriateness of the PCA, was 0.68. This figure exceeds the minimum KMO criteria of 0.50 for PCA analysis. Given the fact that the first principal component explains more than 70% of the variations in the original variable, the eigenvectors of the first principal component was used as weight in the construction of the financial inclusion index. Thus, a higher score implies that financial inclusion is relatively high and a lower score implies lower levels of financial inclusion.

4.4.2 *Time Series Properties and Descriptive Statistics*

Before carrying out the structural VAR analysis, the time series properties of the variables were checked. Results of the unit roots test are presented in Table 4.1 of the Appendix. The study used the Philips Perron (PP) test and the KPSS test to check the stationarity of the series. These two tests are known to have stronger power and tend to be more efficient in identifying stationary and non-stationary series. The results of the PP-test identified inflation rate, Lending rate, and CIEA as stationary at levels and the others only stationary after first difference. On the other hand, the KPSS identified Inflation, MPR, Lending Rate, Deficit, and CIEA as stationary at levels. The only non-stationary variables according to the KPSS were Exchange rate and Financial Inclusion. Since the KPSS test is known to have stronger power of test than the KPSS, decision on stationarity of variable was based on the KPSS. In order to estimate the SVAR model, the first difference of the two non-stationary variables (Exchange Rate and Financial Inclusion) were used in addition to the stationary ones for the SVAR estimation.

Two separate descriptive statistics are presented in Appendix 1. Table 4.2 presents the descriptive statistics for the variables used for the PCA analysis and Table 4.4 presents the descriptive statistics for the variables employed in the SVAR model. From Table 4.2, average savings over the study period was found to be GH¢6,344.53 million. Average outstanding loan balance of banks was 15.7%. Commercial Bank branches and ATMs per 1000km^2 averaged 5.2 and 6.1 respectively. An examination of the minimum and maximum values of OLCB, CBB and ATM seem to suggest that progress towards increasing accessibility and usage of financial services in terms of these variables have progressed steadily. However, a comparison of the standard deviation and mean of savings seem to suggest that saving in the country over the study period has largely been affected by outliers. Hence to reduce the possible effect of outliers on our estimation, the study took the natural log of savings to help normalise the data.

Table 4.4 presents the descriptive statistics for variables employed for the SVAR model. Inflation rate over this period averaged 13.9% with budget deficit also averaging 5.6% of GDP. Average lending rate over this period exceeded the monetary policy rate by close to 10%. Economic activities in the country grown steadily over the period averaging 9.3% per month.

4.5 Results and Discussions

4.5.1 Impulse Response Function

The discussions of the impulse response function focus on the policy objective of the central bank. i.e., inflation and CIEA. This is because monetary policy decisions aim at achieving price stability and helping to achieve sustainable growth rate in the country. Hence, the discussion on the impulse response function is based on how inflation and CIEA respond to changes in monetary policy, financial inclusion and the interaction between monetary policy and financial inclusion. Results of both short run and long run impulse response functions (IRFs) are presented in Fig. 4.1.

Results of the short run IRF indicates that a positive shock to the monetary policy rate leads to increases in the rate of inflation in the short run. That increase in the MPR is associated with increases in the rate of inflation at least within the first six months of the positive shock to monetary policy before inflation begins to decline. This can be partly attributed to the level friction in Ghana's financial sector which does not allow for the smooth transmission of monetary policy in the economy. These frictions in the past have included high level of financial exclusion in the economy and the lack of coordination between monetary policy makers and fiscal policy makers. The initial rise in the rate of inflation can also be attributed to the lack of development of Ghana's financial sector. Taylor (1983) and Van Wijnbergen (1982) maintain that, due to weak equity and commodity markets in several developing countries (including Ghana), firms tend to depend on the credit market to finance their working capital needs, hence increase in the interest rate contributes to increase the cost of production of firms which is transmitted to the prices of goods and services leading to an increase in the overall inflation of the economy. This argument is consistent with results obtained for this study. The reduction in the rate of inflation six months after the shock could be attributed to reduction in demand for goods and services due to the initial price hikes and the high interest rate.

The study found a positive shock to financial inclusion (FIN) associated with reduction in the rate of inflation however this was only significant four months after the shock. That is, increase in the accessibility and usage of financial services contribute to reduction in inflation. Increase in accessibility and usage of financial services imply more people would be included in the financial sector and are more likely to reduce their demand

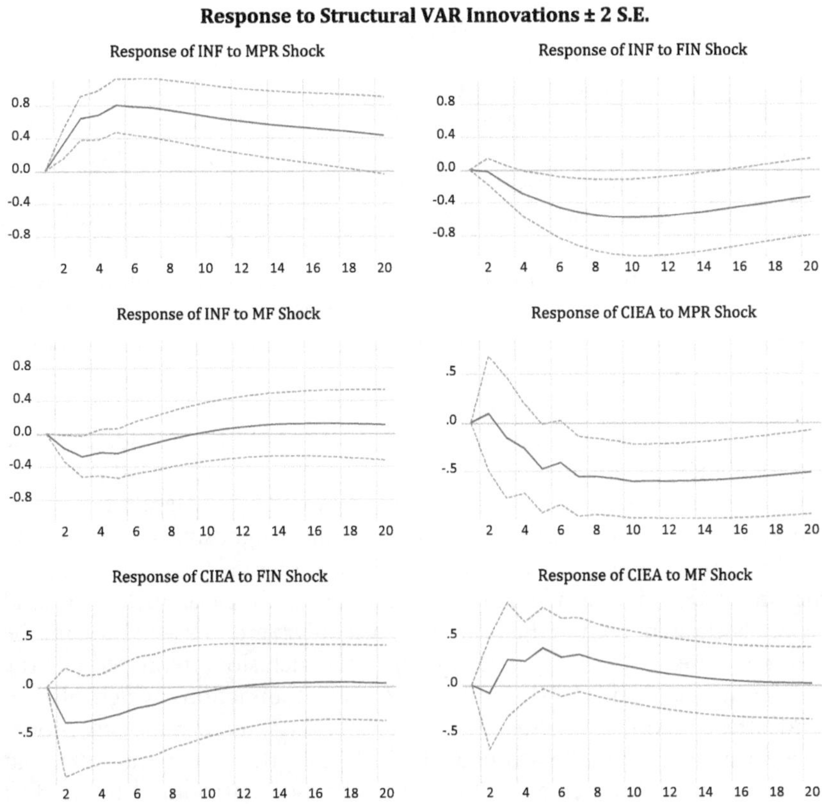

Fig. 4.1 Short run impulse response functions (*Source* Authors' Computations)

for goods and services which would go a long way to reduce general price level. This is because people who are financially included are more likely to save more and invest in interest yielding financial assets, thereby reducing the amount of money available for transitionary purposes which would eventually reduce demand for goods and services and invariably contribute to the reduction in the rate of inflation. The impact of financial inclusion on inflation tends to be immediate when policy makers take the level of inclusion into account (see the response of INF to MF shock in Fig. 4.1). The results imply that increase in the level of financial inclusion and monetary policy could help the Bank of Ghana to effectively achieve its short to medium term targets. This is because an understanding of the level of financial inclusion in the economy could help the central bank to gauge the specific group of people who are likely to be affected and recommend specific policy options that would maximise the expected effect of the policy decisions. This is due to the fact that, increase in the levels of financial inclusion tend to increase the sensitivity of inflation to interest rate (see Brownbridge et al., 2017; Mehrotra & Nadhanael, 2016).

With regards to the impact of monetary policy on economic activity, we found that increase in the monetary policy rate is associated with declines in economic activities as expected. Increase in monetary policy rate tends to reduce the availability of credit for economic activities hence causing reduction in economic activities. The reduction in the economic activity due to increased monetary policy rate could be attributed to the fact that increase in monetary policy rate tends to increase the cost of borrowing through the impact of monetary policy on lending rates in the economy. This in tend reduces the ability of businesses to expand production causing output to decline. In the case of financial inclusion, the study found that increase in the level of financial inclusion tend to be associated with reduction the level of economic activities in the country. This is probably because increase inclusion, tend reduce the monies in the hand of the public for daily transactions which also in tend affects economic activities in the country. However, when monetary policy is interacted with financial inclusion, the results indicates that economic activities increase. This can probably be attributed to the fact that increase in the level of financial inclusion tends to increase the usage of all forms of financial services, hence, whilst increase in the policy rate help mobilise excess liquidity within the economy, the increased inclusion also tends increase the availability of funds for long term investments, which invariably increases economic activities.

In terms of the long-term (See Appendix) response to monetary policy and financial inclusion shocks, the study found that in the long run, an increase in the monetary rate significantly resulted in the reduction in inflation. Similar result is obtained when there is a positive shock to financial inclusion. However, in the case of the financial inclusion, equilibrium is restored faster than the monetary policy shock. When the interaction between monetary policy and financial inclusion is also considered, the study found that there was a gradual increase in inflation from relatively lower levels towards the equilibrium rate. Thus, the results of the three response functions of inflation to a positive shock in monetary policy, financial inclusion and the interaction between financial inclusion and monetary policy reveals that, price stability would be attained rapidly when the level of financial inclusion is high. Similar results are obtained for response of the composite index of economic activities.

4.6 Conclusions and Policy Implications

Whilst the implications of financial inclusion for growth and inequality is well established in the literature, little attention has been given to the implication of financial inclusion for monetary policy. This chapter sets out to examine the effects of financial inclusion on monetary policy in the Ghanaian economy. The study estimated a structure VAR model and found evidence that increases in the level of financial inclusion tend to increase the responsiveness of inflation and output to changes in the monetary policy rate over time. That is when monetary policy rate is interacted with financial inclusion, the study found that it had faster effect in reducing inflation in the short run than the sole implementation of monetary policy. In the long run also, the study identified that price stability is attained faster when monetary policy is interacted with the level of financial inclusion.

Results of the study suggest the need for the Central Bank to strengthen its efforts to improve financial inclusion in Ghana. Since high level of inclusion tend to increase the sensitive of output and inflation to changes in interest rate. Also, there is theneed for Bank of Ghana to consider gathering monthly data on usage and accessibility of financial services in the country. This would help inform the Central Bank on the level of inclusion in the country which would be key in arriving at effective monetary policy decisions to promote the development of the financial sector.

Appendices

See Table 4.1
See Table 4.2
See Table 4.3
See Table 4.4
See Fig. 4.2
See Table 4.5

Table 4.1 Time series property of series used

	PP-test		*KPSS test*	
	Levels	*First difference*	*Levels*	*First difference*
Inflation	−2.7[0.08]	−11.3[0.00]	0.29	0.04
Exchange rate	1.1 [0.99]	−16.7[0.00]	1.68	0.33
MPR	−2.1[0.24]	−14.8[0.00]	0.30	0.24
Lending rate	−3.1[0.03]	−14.0[0.00]	0.57	0.22
Deficit	−2.5[0.11]	−3.93[0.00]	0.41	0.05
CIEA	−5.2[0.00]	−25.6[0.00]	0.48	0.32
Financial inclusion	1.6 [0.99]	−6.13[0.00]	1.75	0.39

Table 4.2 Descriptive statistics: Financial inclusion variables

	OLCB	*CBB*	*ATM*	*SAVINGS_TD*	*LSAVINGS*
Mean	15.7	5.2	6.1	6344.5	8.0
Median	15.8	5.3	3.9	3815.1	8.2
Maximum	19.8	8.8	11.8	25,609.2	10.2
Minimum	10.1	2.6	3.5	295.8	5.7
Std. Dev	2.4	1.8	3.1	6821.6	1.3
Skewness	−0.1	0.4	0.8	1.3	−0.2
Kurtosis	2.3	2.3	1.9	3.6	1.8
Jarque–Bera	3.9	8.9	31.9	57.7	13.1
Probability	0.1	0.0	0.0	0.0	0.0
Sum	3196.4	1065.9	1237.5	1,294,284.0	1640.2
Sum Sq. Dev	1202.3	663.7	1968.9	9,450,000,000.0	358.9
Observations	204.0	204.0	204.0	204.0	204.0

Source Authors' Computation

Table 4.3 PCA results

Component	*Eigenvalue*	*Difference*	*Proportion*	*Cumulative*	*Variable*	*KMO*
Comp1	1.88965	1.89313	0.7024	0.7024	CBB	0.6492
Comp2	1.096521	0.896384	0.2691	0.9715	ATM	0.8533
Comp3	0.100137	0.0864471	0.025	0.9966	OLCB	0.2112
Comp4	0.0136898		0.0034	1	SAVINGS_TD	0.627
					Overall	0.6761

Table 4.4 Descriptive statistics main variables

	Inflation	*Exchange rate*	*MPR*	*Lending rate*	*Deficit*	*CIEA*	*Financial inclusion*
Mean	13.9	1.8	18.4	28.3	−5.5	9.3	0.0
Median	13.1	1.4	18.0	27.5	−5.3	8.3	−0.3
Maximum	29.9	4.5	27.5	40.0	−1.6	29.6	3.2
Minimum	5.6	0.7	12.5	23.7	−12.6	−5.1	−2.2
Std. Dev	5.0	1.2	4.5	3.6	2.2	6.9	1.7
Skewness	1.0	1.2	0.5	1.1	−1.0	0.5	0.5
Kurtosis	4.4	2.9	2.1	3.6	4.8	2.6	2.0
Jarque–Bera	50.8	45.3	16.6	44.2	58.8	9.8	16.0
Probability	0.0	0.0	0.0	0.0	0.0	0.0	0.0
Sum	2825.7	362.2	3755.5	5771.5	−1117.7	1898.8	0.0
Sum Sq. Dev	5105.8	281.5	4047.4	2597.1	982.6	9685.2	572.1
Observations	204	204	204	204	204	204	204

Source Authors' Computation

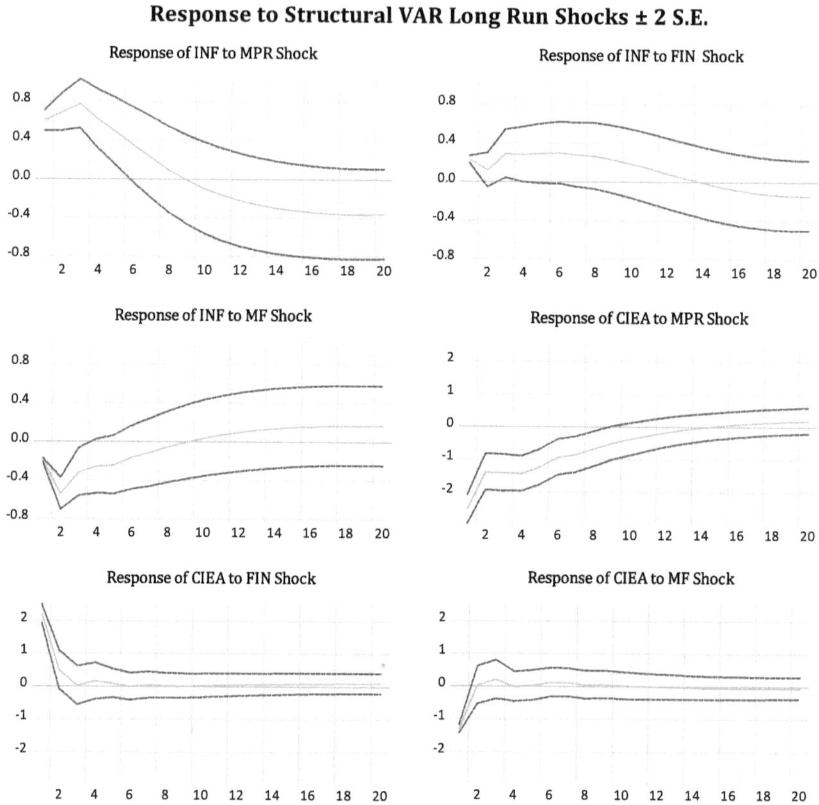

Fig. 4.2 Long run impulse response functions (*Source* Authors' Computations)

Table 4.5 ASVAR pre-estimation tests

Lag selection criterion

Lag	*LogL*	*LR*	*FPE*	*AIC*	*SC*	*HQ*
0	−3469.948	NA	428,585.6	35.67126	35.80554	35.72563
1	−1287.779	4162.906	0.000158	13.94645	15.15495	14.43576
2	−735.5136	1008.239	1.06E-06	8.938601	11.22131	9.862843
3	−526.5264	364.388	2.41E-07	7.451553	10.80848*	8.810731*
4	−464.1831	103.5858	2.48E-07	7.468545	11.89968	9.26266
5	−398.785	103.2954	2.51E-07	7.454205	12.95956	9.683258
6	−337.1446	92.30259	2.67E-07	7.478406	14.05797	10.1424
7	−251.3123	121.4856	2.25e-07*	7.254486	14.90827	10.35341
8	−184.364	89.26450*	2.34E-07	7.224246*	15.95224	10.75811
LM serial correlation test						
Null hypothesis: No serial correlation at lag h						
Lag	LRE* stat	df	Prob	Rao F-stat	df	Prob
1	54.59459	64	0.7928	0.847905	(64, 669.8)	0.7938
2	69.32282	64	0.3027	1.088189	(64, 669.8)	0.3042
3	86.2776	64	0.0332	1.371125	(64, 669.8)	0.0336
4	101.0295	64	0.0022	1.622931	(64, 669.8)	0.0022
5	66.79095	64	0.3813	1.046523	(64, 669.8)	0.3829
6	80.75971	64	0.0768	1.278292	(64, 669.8)	0.0775
7	147.3323	64	0	2.448615	(64, 669.8)	0
8	55.69971	64	0.7606	0.865759	(64, 669.8)	0.7617
9	63.13376	64	0.5071	0.986602	(64, 669.8)	0.5087

References

Anarfo, E. B., Abor, J. Y., Osei, K. A., & Gyeke-Dako, A. (2019). Monetary policy and financial inclusion in Sub-Sahara Africa: A panel VAR approach. *Journal of African Business*, 1–24.

Bernanke, B. S. (1986). Alternative explanations of the money-income correlation. *Carnegie-Rochester Conference Series on Public Policy, 25*, 49–99. http://www.sciencedirect.com/science/article/pii/0167223186900370

Brownbridge, M., Bwire, T., Rubatsimbira, D. K., & Tinyinondi, G. A. (2017). The impact of financial inclusion on the interest rate channel of the monetary policy. *Bank of Uganda Working Paper*, *No. 05*.

Cushman, D. O., & Zha, T. (1997). Identifying monetary policy in a small open economy under flexible exchange rates. *Journal of Monetary Economics, 39*(3), 433–448.

De Bondt, G. J. (2000). *Financial structure and monetary transmission in Europe: A cross-country study*. Edward Elgar Publishing.

Denton, F. T. (1971). Adjustment of monthly or quarterly series to annual totals: An approach based on quadratic minimization. *Journal of the American Statistical Association, 66*(333), 99–102.

Evans, O. (2016). The effectiveness of monetary policy in Africa: Modeling the impact of financial inclusion. *Iranian Economic Review, 20*(3), 327–337.

Hove, S., Touna Mama, A., & Tchana Tchana, F. (2015). Terms of trade shocks and inflation targeting in emerging market economies. *South African Journal of Economics*. http://onlinelibrary.wiley.com/doi/10.1111/saje.12086/pdf

Jolliffe, I. T., Trendafilov, N. T., & Uddin, M. (2003). A modified principal component technique based on the LASSO. *Journal of Computational and Graphical Statistics, 12*(3), 531–547.

Lapukeni, A. F. (2015). The impact of financial inclusion on monetary policy effectiveness: The case of Malawi. *International Journal of Monetary Economics and Finance, 8*(4), 360–384.

Le, T.-H., Chuc, A. T., & Taghizadeh-Hesary, F. (2019). Financial inclusion and its impact on financial efficiency and sustainability: Empirical evidence from Asia. *Borsa Istanbul Review, 19*(4), 310–322.

Lenka, S. K., & Bairwa, A. K. (2016). Does financial inclusion affect monetary policy in SAARC countries? *Cogent Economics & Finance, 4*(1), 1127011.

Mbutor, M. O., & Uba, I. A. (2013). The impact of financial inclusion on monetary policy in Nigeria. *Journal of Economics and International Finance, 5*(8), 318–326.

Mehrotra, A., & Nadhanael, G. V. (2016). Financial inclusion and monetary policy in emerging Asia. In *Financial inclusion in Asia* (pp. 93–127). Springer.

Peprah, J. A., Koomson, I., Sebu, J., & Bukari, C. (2021). Improving productivity among smallholder farmers in Ghana: Does financial inclusion matter?

Agricultural Finance Review, 81(4), 481–502. https://doi.org/10.1108/AFR-12-2019-0132

Samba, M. C. (2018). *Monetary policy effectiveness under the CEMAC area: An empirical evaluation. IJAME.*

Sims, C. A. (1986). Are forecasting models usable for policy analysis? *Federal Reserve Bank of Minneapolis Quarterly Review, 10*(1), 2–16.

Taylor, L. (1983). *Structuralist macroeconomics: Applicable models for the third world.* Basic Books.

Van Wijnbergen, S. (1982). Stagflationary effects of monetary stabilization policies: A quantitative analysis of South Korea. *Journal of Development EconomIcs, 10*(2), 133–169.

Woodford, M. (2001). The Taylor rule and optimal monetary policy. *The American Economic Review, 91*(2), 232–237.

CHAPTER 5

Gendered Financial Behaviour in Ghana: A Comparative Study with South Africa

Shoba Arun, Samuel Annim, Udichibarna Bose, and Thankom Arun

5.1 Introduction

Women's integration into global finance has been gaining much focus in recent efforts towards financial inclusion (defined as the delivery of financial services at affordable costs to disadvantaged groups) debates. As a key enabler for development, financial inclusion is firmly placed on the agenda of most governments as a key policy priority (Arun & Kamath, 2015; Global Findex, 2015). However, globally the gender gap

An earlier version of this chapter was published under the title 'Even' After Access to Financial Services? Ricocheting Gender Equations' as an IZA Discussion Paper No. 10099. However, this version of the chapter is substantially different from the previous one in terms of the narrative, background literature and discussion of the empirical results.

S. Arun
Department of Sociology, Manchester Metropolitan University, Manchester, UK
e-mail: s.arun@mmu.ac.uk

J. A. Peprah et al. (eds.), *Financial Sector Development in Ghana*, Palgrave Macmillan Studies in Banking and Financial Institutions, https://doi.org/10.1007/978-3-031-09345-6_5

in finance, referring to the disproportionate exclusion of women from access to and usage of formal financial services, is seen to be essentially static, with a steady 9 percentage point gap over the period 2011–2014 (Lewis et al., 2016). This points to enhanced efforts to enable women to gain more control over finance for invividual and household welfare. Exploring the role of gender in financial services sheds light on the ambiguities of the neoliberal gender agenda in the development discourse. For example, Calkin (2015) has pointed how women and girls are the public faces of anti-poverty policy, while Chant and Sweetman (2012) critique the instrumental policies of 'smart economics' that 'fix' gender rather than the economy, through investing in women for global growth. Burchi and Vicari (2014) also confirm fostering gender equality through people's capability to participate in household decision-making and on intra-household gender relations.

In Africa, the banking system, which forms the bulk of the financial system, has undergone substantial changes over the past two decades (Asuming et al., 2019). Beck et al. (2015) examine the factors affecting financial inclusion in Africa and find that foreign banks from emerging markets helped improve access to finance in African countries. Zins and Weill (2016) examine some determinants of financial inclusion in 37 African countries, and find that being a man, richer, more educated and older is associated with greater financial inclusion in African countries. Allen et al. (2014) show that innovative financial services helped to overcome infrastructural problems and improved access to finance. Evans (2018) examine the relationship between internet, mobile phones and financial inclusion in Africa from 2000 to 2016 and find that the internet and mobile phones improved the ability of individuals to access basic financial services. However, Chikalipah (2017) found that illiteracy is a

S. Annim
Department of Applied Economics, University of Cape Coast, Cape Coast, Ghana

U. Bose · T. Arun (✉)
Essex Business School, University of Essex, Essex, UK
e-mail: t.g.arun@essex.ac.uk

U. Bose
e-mail: ubose@essex.ac.uk

major hindrance to financial inclusion in Sub-Saharan Africa. Thus, the last decade has witnessed renewed efforts by policymakers and the international development community to expand access to affordable financial services to those excluded from the formal financial sector. However, despite these improvements, the formal financial system in Africa is still not very inclusive (Beck et al., 2015).

This chapter contributes to the literature on financial inclusion by focusing on two sub-Saharan African countries, namely Ghana and South Africa, in relation to how gender is a crucial part of the mutually constitutive process within financial services. The main contributions of this study are mainly threefold. First, we undertake a comparative analysis between Ghana and South Africa using the FinScope Data to determine the extent to which gender differences affect the use of different financial products such as general accounts and services, insurance and investment services. Second, we study gender bias in using higher order financial products by classifying each financial product into four categories based on their level of complexity. Finally, we study the extent to which other observed demand and supply side characteristics influence the use of different financial products.

This chapter is divided into five sections. In section two, we introduce the context of gender equality and financial sector policies in Ghana and South Africa, followed by a detailed review of the literature on gendered aspects of economic and financial behaviour. Section three outlines the data and empirical methodology used in this study, while section four discusses the empirical findings. Finally, section five concludes the chapter.

5.2 Background Literature

5.2.1 The Context of Ghana and South Africa

This chapter focuses on Ghana with a comparative study with South Africa based on converging and diverging economic and financial sectors, and gender equality. At the macroeconomic level, Ghana has exhibited converging trends with South Africa in terms of growth patterns with increasing economic growth and reduction in poverty, and has stark patterns of socio-economic inequality (Annim et al., 2012). Ghana has made strides in reducing poverty to less than 30% of the population and acquiring the lower middle-income status, while South Africa is an upper middle income country but is characterised by multiple forms of inequality (Barbarin & Richter, 2013; UNDP, 2014). Women's unequal status in Ghana and South Africa is reflected through various socio-economic disparities (Deghaye et al., 2014; Heintz, 2005) and efforts to address gender inequalities in post-apartheid era (Albertyn, 2011). Further, Morsy (2020) finds that women are self-selecting out of the African credit market.

With issues of women's financial and economic security disproportionately represented in the labour and credit markets, the similarities and differences to socio-economic patterns of inequality have implications on financial inclusion particularly for women and form an important background to this study. It also renders significant comparisons within these two countries per se not just for exploring gendered access to resources, but also to locate the nature of these in relation to gender equality and financial inclusion. Compared with other Sub-Saharan countries and Nigeria, Angel-Urdinola and Wodon (2010) find that most household decisions are made by men, who are the de facto household heads. While, women take part more often in decisions on expenditures for food, heath and education, but even in these areas, men more often than not remain the main decision makers. Byrakatar and Fofack (2018) present an Overlapping Generations Model of economic development for gender and growth analysis in a low-income country environment. In addition to the formal sector the model considers informal production, which accounts for an increasingly large share of aggregated output in Sub-Saharan Africa in the absence of barriers to entry. The model also accounts for financial market imperfections and its disproportionately larger adverse effects on women's access to credit. Policy experiments derived from an application of the model to Burkina Faso yield exciting results. In particular, a reduction in the cost of borrowing, increased women's bargaining power and higher public spending on education are growth-enhancing. Empirical evidence from the model shows that these policies are also welfare-enhancing and growth-inclusive. They tend to narrow gender gaps in economic well-being and strengthen women's economic empowerment.

Although Ghana and South Africa exhibit some divergences in terms of regulation and policies in the financial sector, they do present higher levels of financial exclusion as in many developing countries. As part of the financial sector reform that started in 1987, Ghana abolished interest rate caps and credit allocation to priority sectors. Recent legislation on venture capital (2004), insolvency (2006) and credit reporting (2006) shows the desire to create an enabling atmosphere in Ghana. Further, changes in the regulatory framework ensure significant improvements in the legal basis for financial intermediation particularly in prudential supervision, capital adequacy, bank risk management, and more on-site supervision. While in South Africa, we have observed intermittent downturns in access and use of financial services, although it enjoys better access to finance than other countries in the region. For instance, the FinMark Trust (2011) reports a 3% drop in the proportion of use of bank services by South Africans, and a decline in the demand-side factors influencing access to financial markets (Kostov et al., 2011). The financial environment in South Africa has also witnessed substantial changes in financial regulations and innovations, and partnerships across all financial institutions (Annim et al., 2012), such as the Mzansi initiative introduced in 2004 to reach the 'unbanked'.

While it is apparent that in both Ghana and South Africa conscious efforts have been instituted to ensure access to financial services, this study examines whether this automatically translates to the use of financial services. For the purposes of this chapter, the types of financial services relate to basic bank accounts and services, insurance services and investment services.

5.2.2 *Gender and Financial Use: Through the Lens of Ambivalence*

By now, there is an exhaustive body of scholarship on women's access to finance, with significant attention on supply side factors related to gendered access to different forms of finance. Much of this debate is located within the entrepreneurship (and gender) literature as to how gender mediates external business finance (Marlow & Swail, 2015) or economic psychological evidence in finance and risk taking (Meier-Pesti & Penz, 2008) and access to micro finance in the global south (Maclean, 2010). The research has primarily focused on attitudes towards financial risk in economics, and the evidence on gender differences is more mixed. Studies have shown that the economic well-being and financial behaviours of men and women differ significantly. Researchers have reported that women invest their financial resources more conservatively and are more risk averse than men (Charness & Geezy, 2012). Women have lower rates of participation in retirement plans as compared with men and hence, are more likely to be living in poverty during retirement.

Eckel and Grossman (2002) survey the economics literature through the economic psychology lens, comparing the data across abstract gambles, contextual experiments and field studies to show that the results from field studies point to how women are more risk averse. However, the findings of laboratory experiments are less clear. Feminist perspectives on decision-making power have discussed the differences in economic behaviours within the household, exploring power relations and as critical sites of decision-making and distributional inequality. The vast scholarship on the critique of unitary models of households that operates based on economic rationality points to the asymmetric intra-household bargaining power and allocation of resources based on gender (Folbre, 1986; Katz, 1997; McElroy & Horney, 1981). Evidence on intra-household inequality within the Global South points towards non-material and material foundations of bargaining power (See Doss, 2013). For Sen (1990), determinants such as female labour-force participation and literacy are key, while Agarwal (1997) has highlighted the role of social norms and socially recognised and sanctioned gender roles in bargaining. The intra-household distribution of control over resources and power affect both intra-household decision-making processes, and economic or financial decision-making in other spheres.

Drawing on the existing literature on the determinants of intra-household bargaining and decision-making power, Ashraf (2009) elicit causal effects of spousal observability and communication on financial choices of married individuals in the Philippines. Further, Aterido et al. (2011) use individual-level survey data for nine countries in Sub-Saharan Africa to examine the gender gap in financial services. They find that women's lower use of formal financial services in nine Sub-Saharan Africa countries can be explained by gender differences in education and income levels, formal employment and being the head of household.

As seen from the economics literature, observed gender differences in demand for finance and credit are a consequence of choice, outcome of systemic disadvantage or financial discrimination (van Hulten, 2012). Morsy and Youssef (2017) study the impediments women face, compared with men, when trying to access finance and suggest that women are more likely to be excluded from the formal financial sector in countries where laws and norms discriminate against women. Further, Arnold and Gammage (2019) provide an extensive survey on gender and financial inclusion and deliver insights that enable practitioners to better reach women with digital financial services. Sociological approaches, primarily Bourdieu's (1977, 1984) unique concept of habitus, also helps shed light on social behaviours. Habitus theorises a self that is socially produced and constitutive of social relations, and is a 'socialised subjectivity' (Bourdieu & Wacquant, 1992), where by one's practice is interlocked with one's habitus, relative capital, and with the field in which one operates. For this chapter, the notion of habitus referred to in *Distinction*, where, 'Being the product of the conditionings associated with a particular class of conditions of existence, [the habitus] unites all those who are the product of similar conditions while distinguishing them from all others....' (Bourdieu, 1984). The concept of gendered habitus and intra-household decision-making helps unravel how practices of early socialisation and social norms affect behaviour and practices in later stages, such as financial decision-making. An emerging body of scholarship points to how global finance is gendered through financial markets, often considered a male-dominated sphere characterised by risk taking, speculation and profits (Assassi, 2009). Thus, this chapter explores more evidence on the practices of financial behaviour and norms of financial structures inflected by a range of gendered and social behaviours in the demand for and use of financial services in Ghana and South Africa.

5.3 Data and Empirical Model

This chapter uses data from the FinScope surveys in Ghana and South Africa in 2010, an initiative of the FinMark Trust (an independent trust funded primarily by UKaid, the United Kingdom's Department for International Development). The survey is nationally representative and conducted in a number of Sub-Saharan African countries, on several issues including financial services and socio-economic characteristics. The survey, drawn from the individual adult population (above 15 and 16 years in Ghana and South Africa, respectively), is based on a multi-stage sampling approach where 3648 respondents were interviewed in Ghana and 3900 in South Africa. Some country-specific issues were incorporated, for example, an important element of the South African financial market is the introduction of Mzansi accounts. Thus, the analysis takes this into account by classifying access to Mzansi account as part of the 'none' category. The content of the FinScope instrument is as follows: (i) Household register; (ii) Financial literacy; (iii) Overall financial perception; (iv) Banking penetration (transaction channels, Mzansi and credit and loans); (v) Insurance products and services (funeral cover and retirement/pension); (vi) Investment/savings; (vii) Lifestyles; (viii) Access to amenities and use of information, communication and technology (ICT);

(ix) Sources of money and (x) Personal and household's socio-economic and demographic characteristics (Arun et al., 2016).

For the empirical analysis of this chapter, we first estimate an ordered logit model for the three types of financial products and then use a non-linear version of the Blinder-Oaxaca decomposition technique that decomposes the coefficient ascertained from the ordered logit model into two parts. Three outcome variables (types of financial services used) were constructed based on different financial products: (i) level of use of general accounts and services, (ii) level of use of insurance services and (iii) level of use of investment services. Each of the financial products were further classified into levels of use, namely: pre-account (none) (coded as 1); basic account (coded as 2); intermediate account (coded as 3) and advanced account (coded as 4). The gender of respondents was assigned as 1 for females and 0 for males. We also include variables such as age, income level, education and household size of the respondents and control for regional and provincial effects.

We also include supply side factors that affect the decision to use financial services, and these include: working hours of financial institutions; cost of service provision (interest rate and other transaction costs); return on deposits; staff attitude; travelling distance to financial institutions; time spent in engaging in a financial service; and a sense of intimidation in the banking hall. These variables were captured by counting the number of such constraints faced by each individual. With Ghana, we identified eight such questions and five in South Africa. Table 5.1 presents how the various services were categorised.

The choice of ordered logit is appropriate because the dependent variables are ranked from 1 to 4 in each case, with none or pre-account coded as 1, basic account as 2, intermediate account as 3, and advanced account as 4. We then estimate the odds ratios since the coefficients in themselves mean very little in terms of interpretation. To decompose the coefficient associated with 'gender', based on observed and unobserved characteristics of men and women, we employ the Blinder-Oaxaca decomposition for non-linear regression models (Sinning et al., 2008). This procedure is used instead of the usual Blinder–Oaxaca decomposition because it allows us to decompose the outcome variable of a non-linear dependent variable into a part that is explained by observed characteristics and part attributable to estimated coefficients (Arun et al., 2016).

Table 5.1 Classification of type and levels of use of financial service

Types of financial service	*Levels*			
	Pre (None)	*Basic*	*Intermediate*	*Advance*
General accounts and services	Mzansi and loan of a friend	ATM, savings book, post office account, savings and transaction account, employer or microcredit institution	Debit card, current or cheque account, credit card, fixed deposit, mortgage, money for house either from government or employer and personal loan	Money market, vehicle finance or overdraft
Investment	None	Lending to others and benefitting from their profit and being a member of Stokvel/savings club or burial society	Investment in property/house/flat that you rent out, Unit trusts, Investment in cattle/livestock and collectables (antiques/carpets/paintings/art/coins/stamps)	Timeshares, holiday home/investment in a second home, investment in vacant land, farm land, own business, someone else's business, shares on the stock market and off-shore investment
Insurance	None	Funeral policy with a big institution, provident or pension fund and educational policy	Life assurance policy, retirement annuity, endowment/investment policy, homeowner's insurance, medical aid, insurance taken out to maintain credit payments and car insurance	Hospital plan and medical or household content insurance

Given an ordered logit regression model regression model for two groups $g = (m, f)$, because the outcome variable is non-linear the conditional expectation of $E(Y_{ig}|X_{ig})$, may differ from $\overline{\boldsymbol{X}}_g\widehat{\beta}_g$ (Sinning et al., 2008). Sinning et al. (2008) obtains a general version of the Blinder Blinder-Oaxaca decomposition as:

$$\Delta_m^{NL} = \left[E_{\beta_m}(Y_{iA}|\boldsymbol{X}_{im}) - E_{\beta m}(Y_{if}|\boldsymbol{X}_{if})\right] + \left[E_{\beta m}(Y_{if}|\boldsymbol{X}_{if}) - E_{\beta f}(Y_{if}|\boldsymbol{X}_{if})\right]$$

where $E_{\beta_g}(Y_{ig}|\boldsymbol{X}_{ig})$ is the conditional expectation of Y_{ig} and $E_{\beta_g}(Y_{ih}|\boldsymbol{X}_{ih})$ represents the conditional expectation of Y_{ih} evaluated at the parameter vector β_g. In this set up $g, h = (m, f)$ and $g \neq h$. . The first term on the right hand side displays the differential in the outcome variables (in this case general account and services, insurance and investment services) between m and f as a result of differences in the explanatory variables in $\boldsymbol{X}_{ig}$, and the other term shows the part differential that is due to differences of the coefficient. Given the above general procedure we estimate the ologit version of the techniques as:

$$\frac{1}{N_g}\sum_{1=1}^{N}\Big[\Big\{\Lambda\Big(\widehat{\mu_1} - \boldsymbol{X}_{ig}\widehat{\beta_g}\Big) - \Lambda\Big(-\boldsymbol{X}_{ig}\widehat{\beta_g}\Big)\Big\} + 2\Big\{\Lambda\Big(\widehat{\mu_2} - \boldsymbol{X}_{ig}\widehat{\beta_g}\Big)$$
$$-\Lambda\Big(\widehat{\mu_1} - \boldsymbol{X}_{ig}\widehat{\beta_g}\Big)\Big\} + \ldots J\Big\{1 - \Lambda\Big(\widehat{\mu_{j-1}} - \boldsymbol{X}_{ig}\widehat{\beta_g}\Big)\Big\}\Big]$$

where J is the number of possible outcomes and $\widehat{\mu}, \ldots \mu_{J-1}$ are the estimated threshold values of the ordered logit and a is the cumulative logistic density function.

5.4 Findings and Discussion

The results from the analysis are preceded by a descriptive overview of the use of financial services across the different types and the levels of each type. Figures 5.1, 5.2, 5.3 present the descriptive results based on gender differences between Ghana and South Africa. Figure 5.1 shows that more than half of both men and women in Ghana either do not use a general account or service or use a pre-account. We observed the same pattern for the use of investment products (Fig. 5.2) and insurance services (Fig. 5.3). Further, Tables 5.2 and 5.3 present the variables' distribution across gender, showing significant differences between men and women in Ghana and South Africa.

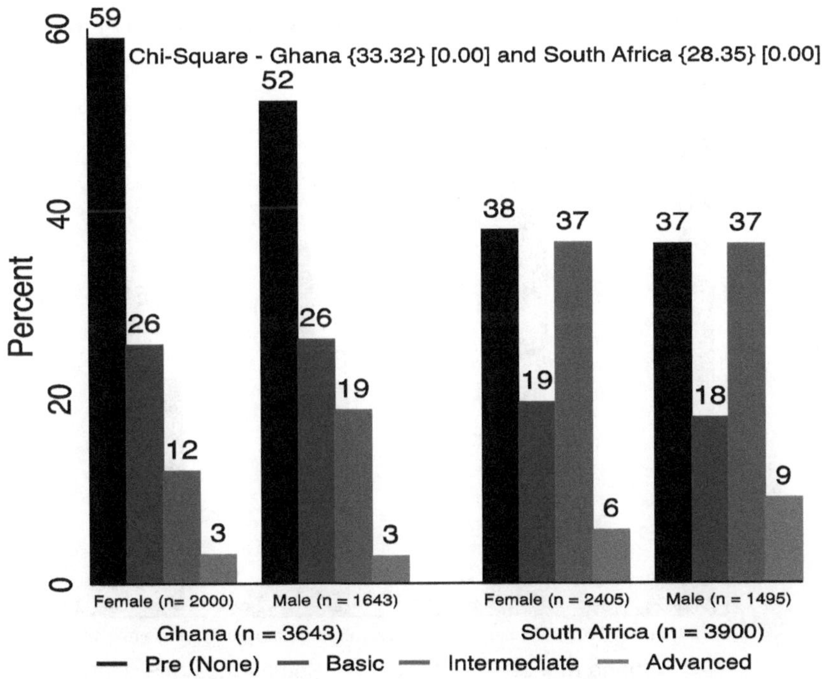

Fig. 5.1 Levels of use of general accounts and services by gender in Ghana and South Africa in 2010

Figure 5.1 gives the histogram for the use of general accounts and services between Ghana and South Africa. We observe larger gender differences in favour of men in terms of the use of general financial accounts in Ghana as compared to South Africa. Figure 5.2 shows that in Ghana every one out of ten people use an investment product, while half of both men and women use an investment product in South Africa. Hence, we observe a significant difference in the proportion of use of basic investment products between Ghana and South Africa. Further, the graphs show that women in both countries use basic investment products, while men in both countries use advanced investment products. Next, Fig. 5.3 presents evidence for the gendered use of insurance services in Ghana and South Africa. The graph shows that insurance service is used by only 5% of the female and 11% of male adult population in Ghana, as compared to half of the adult population in South Africa. Also, Tables 5.2

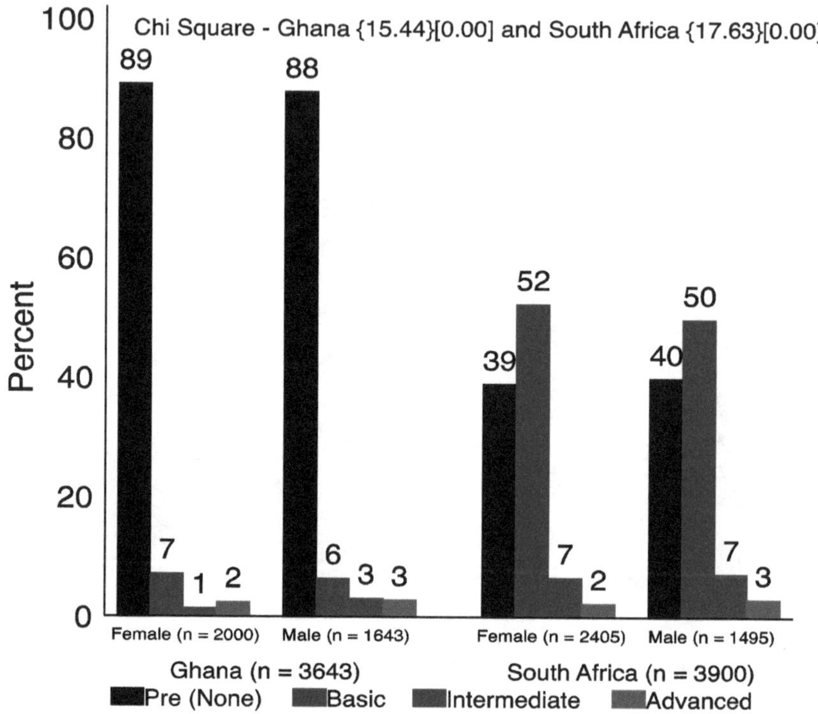

Fig. 5.2 Levels of use of investment products by gender in Ghana and South Africa in 2010

and 5.3 (chi-square tests) show larger gender differences in favour of men in terms of the use of general financial accounts, investments and insurance in Ghana compared to South Africa (Arun et al., 2016). This shows that women's attitude to financial services is associated with gender socialisation, which may not be surprising as evidence shows that many forms of financial behaviour are traditionally associated with the masculine sphere (Marlow & Swail, 2014), and financial markets are constructed as a rational masculinized and professional activity (Assassi, 2009; Doyle & Paludi, 1991).

Tables 5.4 and 5.5 present the ordered logit results for each of the three financial services in Ghana and South Africa. The two tables show that gender significantly influences the use of general accounts and

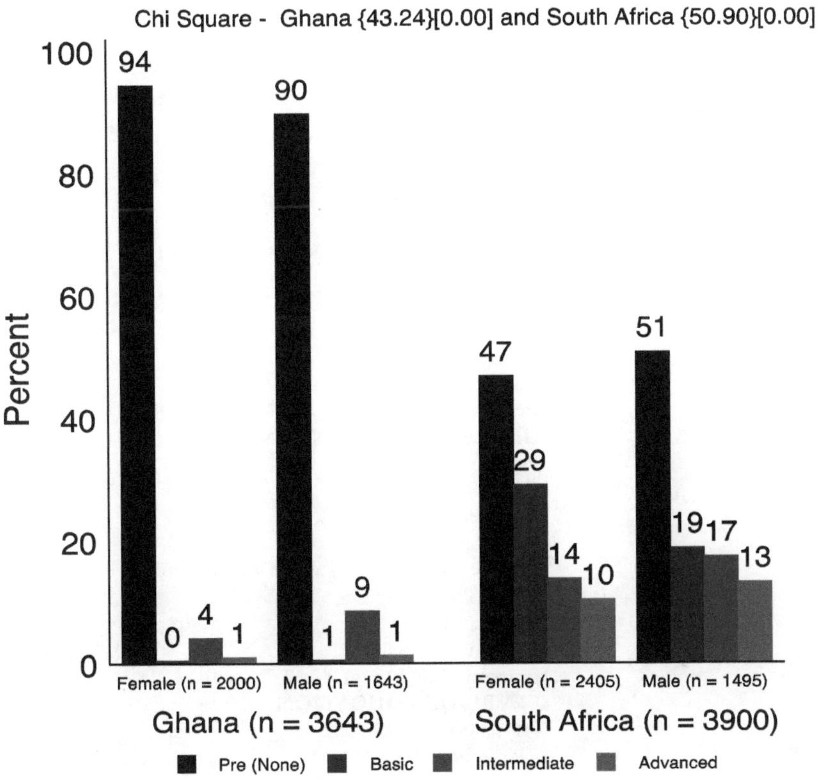

Fig. 5.3 Levels of use of insurance products by gender in Ghana and South Africa in 2010

services and insurance in both Ghana and South Africa (Arun et al., 2016). However, the estimated results fail to find evidence of differences in the use of investment services across gender in either country. Comparing the results in Tables 5.4 and 5.5, we find consistent results for *females* in line with our a priori expectations. According to the results, women in Ghana are 28% less likely to use higher level general account and services, and 43% less likely to use higher level insurance services. However, results in Table 5.5 show that women in South Africa are 26% more likely to use general account and 54% more likely to use insurance services than their male counterparts. Thus, the results show that women in Ghana are yet to be integrated within the financial systems

Table 5.2 Summary statistics of variables-Ghana model

	Male	*Female*
	45%	**55%**
INVESTMENT		
Pre	1211	1506
Basic	86	117
Intermediate	46	26
Advanced	42	38
Pearson chi2(3) = 12.9558 Pr = 0.005		
INSURANCE		
Pre	1228	1603
Basic	7	7
Intermediate	137	65
Advanced	13	12
Pearson chi2(3) = 46.1337 Pr = 0.000		
GENERAL ACCOUNT AND SERVICES		
Pre	732	1001
Basic	336	446
Intermediate	275	195
Advanced	42	45
Pearson chi2(3) = 41.6622 Pr = 0.000		
MARITAL STATUS		
Not married	608	777
Married	777	910
Pearson chi2(1) = 1.4322 Pr = 0.231		
LEVEL OF EDUCATION		
Pre school	303	451
Basic school	665	923
Secondary sch	265	239
Post-secondary	52	26
University	100	48
Pearson chi2(4) = 70.2355 Pr = 0.000		
INCOME LEVEL		
No income	179	303
Below GHC600 per month	1143	1356
GHC601-1200 per month	45	23
GHC1201-1800 per month	11	4
Over 1801	7	1
Pearson chi2(4) = 35.5948 Pr = 0.000		
Total	1385	1687

Table 5.3 Summary statistics of variables-South Africa model

	Male	*Female*
	36%	**63%**
INVESTMENT		
Pre	550	906
Basic	234	590
Intermediate	181	278
Advance	170	214
Pearson chi2(3) = 36.0995 Pr = 0.000		
INSURANCE		
Pre	497	845
Basic	529	999
Intermediate	72	111
Advance	37	33
Pearson chi2(3) = 11.2012 Pr = 0.011		
GENERAL ACCOUNT AND SERVICES		
Pre	386	703
Basic	217	399
Intermediate	422	759
Advance	110	127
Total	1135	1988
Pearson chi2(3) = 11.2903 Pr = 0.010		
MARITAL STATUS		
Not married	703	1237
Married	432	751
Total	1135	1988
Pearson chi2(1) = 0.0249 Pr = 0.875		
INCOME LEVEL		
No income	190	360
Below average	203	318
Average	70	347
Above average	347	621
High income	325	342
Total	1135	1988
Pearson chi2(4) = 115.5599 Pr = 0.000		
LEVEL OF EDUCATION		
Primary Sch	165	375
Secondary sch	403	745
Technical sch	40	27

(continued)

Table 5.3 (continued)

	Male	*Female*
Matriculation and some univer	527	841
Pearson chi2(3) = 27.1918 Pr = 0.000		
Total	1135	1988

as compared to the women in South Africa. This shows heterogeneity among women, reflecting a range of gendered financial behaviours, from different geographical contexts.

Next, we observe that *age* is a significant determinant of the use of all the products in Ghana. It is significant only in explaining the use of insurance services in South Africa. These findings are consistent with earlier works by Annim et al. (2012) and Barslund and Tarp (2008). We find that *education* is also highly significant in explaining the levels of use of financial services in both Ghana and South Africa. These results show that educated people are more likely to use higher levels of these financial services.[1]

Further, rural–urban spatial divides also influence the nature of financial behaviour. A dummy variable (*urban*) was introduced to capture this spatial divide. The results show that residing in an urban area proves to increase the chance of using higher levels of all the financial services, including general account and investment services in Ghana and South Africa. Regional inequalities are also found, where the Western Cape is used as the reference province. In Ghana, compared to the Upper West region, inhabitants of the *Central, Greater Accra, Eastern, Ashanti and Brong Ahafo* regions are more likely to use general accounting services. To examine the effect of being married on the use of these financial services, a *marriage dummy* was introduced in the model, which is positive and significant in some of the models. This shows that social and caring responsibilities and perceptions of ensuing risks result in increased usage

[1] Some studies do provide contrary evidence that professional female investors are inclined to be more risk-averse than male fund managers (Bliss & Potter, 2002; Olsen & Cox, 2001), pointing to the gendered nature of risk aversion even with higher levels of human and educational qualifications. However, while gender differences appear to influence perceptions of risk and recommendations to clients, it is important to note that these differences are the most significant for assets and portfolios at risk extremes.

Table 5.4 Ordered logit estimation of usage of various financial services in Ghana

	(1)		(2)		(3)	
	General account and service		*Insurance*		*Investment*	
	Coef	% Δ	*Coef*	% Δ	*Coef*	% Δ
Female	−0.325***	−27.74	−0.560***	−42.90	−0.0181	−1.790
	(−4.16)		(−3.64)		(−0.15)	
Age	0.0270***	2.734	0.0341***	3.473	0.0169***	1.705
	(10.27)		(7.10)		(4.18)	
HH. Size	0.0248	2.510	0.00113	0.113	−0.0414	−4.054
	(1.07)		(0.03)		(−1.12)	
Urban	0.782***	118.7	0.845***	132.7	0.644***	90.50
	(8.85)		(4.62)		(4.72)	
No. Of reasons	−0.506***	−39.70	−0.259	−22.84	−0.0897	−8.581
	(−6.01)		(−1.65)		(−0.83)	
Married	0.320***		0.208		0.288*	33.4
	(4.02)		(1.32)		(2.31)	
Education (Base = No education)						
Basic sch	0.923***	151.6	0.643*	90.19	0.594**	81.16
	(7.54)		(2.35)		(3.04)	
Secondary sch	1.687***	440.4	1.369***	293.0	0.891***	143.7
	(11.61)		(4.62)		(3.94)	
Post sec	2.609***	1258.8	2.452***	1061.2	1.531***	362.1
	(10.98)		(6.90)		(4.89)	
University	2.959***	1828.7	2.712***	1406.2	2.583***	1224.1
Income (Base = No income)						
Below 600	1.016***	176.1	1.178***	224.8	0.865***	137.6
	(8.27)		(3.82)		(3.95)	
Betw. 601 and 1200	1.286***	261.8	1.519**	356.8	1.311***	270.8
	(4.65)		(3.25)		(3.38)	
Betw. 1201 and 1800	1.400*	305.6	−0.430	−34.9	0.720	105.4
	(2.53)		(−0.28)		(0.94)	
Over 1801	1.132	210.2	0.0902	9.4	0.727	106.9
	(1.67)		(0.09)		(0.76)	
Region (Base = Greater Accra)						
Western	−0.110	−10.45	−0.681*	−49.38	0.0176	1.776

(continued)

Table 5.4 (continued)

	(1)		*(2)*		*(3)*	
	General account and service		*Insurance*		*Investment*	
	Coef	% Δ	*Coef*	% Δ	*Coef*	% Δ
	(−0.72)		(−2.26)		(0.07)	
Central	0.352*	42.17	−0.870*	−58.10	0.00187	0.187
	(2.18)		(−2.46)		(0.01)	
Volta	−0.239	−21.27	0.0696	7.213	−0.0603	−5.848
	(−1.38)		(0.24)		(−0.20)	
Eastern	0.493***	63.72	−0.373	−31.11	0.246	27.84
	(3.35)		(−1.31)		(0.99)	
Ashanti	0.424***	52.74	−0.204	−18.49	0.984***	167.4
	(3.53)		(−0.97)		(5.46)	
Brong Ahafo	0.152	16.47	−1.389**	−75.06	−0.538	−41.61
	(0.89)		(−2.81)		(−1.47)	
Northern	−0.976***	−62.30	−0.709*	−50.81	0.548*	72.97
	(−4.90)		(−1.98)		(2.02)	
Upper East	−0.352	−29.69	−0.205	−18.57	0.823*	127.8
	(−1.46)		(−0.46)		(2.42)	
Upper West	−0.650	−47.79	−0.150	−13.96	0.586	79.77
	(−1.95)		(−0.26)		(1.30)	
cut1						
Constant	3.443***		5.841***		4.851***	
cut2						
Constant	5.095***		5.922***		5.925***	
cut3						
Constant	7.393***		8.304***		6.626***	
N	3072		3072		3072	

Statistical significance is denoted at (*)10%, (**)5% , and (***)1%

of financial services among women, particularly in taking up insurance products, and in line with our observations on social risks and gendered financial behaviour.

In terms of ***income***, we observe that individuals with higher income levels compared with those with no income consistently show a positive and significant effect on the various financial services across all models in both countries. We interpret the finding of a positive and significant effect of income on the use of financial services because of the potential bi-causal relationship. This may also mean that increasing women's income

Table 5.5 Ordered logit estimation of usage of various financial services in South Africa

	(1)		*(2)*		*(3)*	
	General account and services		*Insurance*		*Investment*	
	Coef	%Δ	*Coef*	%Δ	*Coef*	%Δ
Female	0.23*	25.80	0.43***	54.36	0.02	1.72
	(2.15)		(4.12)		(0.16)	
Age	0.00	0.237	0.03***	3.461	0.00	0.17
	(0.56)		(7.91)		(0.41)	
HH. size	−0.01	−0.737	−0.02	−1.914	0.01	0.56
	(−0.31)		(−0.75)		(0.26)	
Urban	0.405**	49.99	−0.15	−14.16	−0.21	−18.79
	(3.10)		(−1.18)		(−1.66)	
No. of reason	0.01	0.726	−0.195	−17.71	−0.06	−5.90
	(0.06)		(−1.81)		(−0.59)	
Married	0.66***	94.30	0.64***	89.70	0.06	5.70
	(5.94)		(5.80)		(0.49)	
Education						
High sch	0.85***	132.90	0.51**	66.51	0.05	5.10
	(4.98)		(3.11)		(0.31)	
Technical sch	2.13***	741.70	1.35***	286.70	−0.37	−30.98
	(5.79)		(3.30)		(−0.96)	
Matriculation/Univ	1.55***	372.70	1.11***	202.10	−0.15	−14.01
	(8.76)		(5.83)		(−0.85)	
Income (No income)						
Below Average	0.63**	87.23	0.40*	49.44	−0.23	−20.79
	(3.20)		(2.07)		(−1.44)	
Average Inc	1.36***	289.20	0.77***	116.0	0.05	5.540
	(6.64)		(3.87)		(0.33)	
Above Av. Inc	1.79***	501.40	1.57***	381.60	0.16	17.41
	(9.41)		(8.46)		(1.02)	
High inc	3.80***	4352.40	4.20***	6566.20	0.44*	55.94
	(14.89)		(17.65)		(2.18)	
Province						
Eastern Cape	−0.34	−28.89	−0.34	−28.46	−0.01	−0.831
	(−1.93)		(−1.72)		(−0.04)	
Northern Cape	−0.38	−31.67	−0.34	−28.51	−1.50***	−77.60
	(−1.86)		(−1.56)		(−6.00)	

(continued)

Table 5.5 (continued)

	(1)		*(2)*		*(3)*	
	General account and services		*Insurance*		*Investment*	
	Coef	*%Δ*	*Coef*	*%Δ*	*Coef*	*%Δ*
Free State	−0.36	−30.45	−0.33	−27.87	0.39*	48.19
	(−1.89)		(−1.47)		(2.26)	
KwaZulu-Natal	−0.18	−16.55	−0.44*	−35.87	0.16	17.31
	(−1.01)		(−2.28)		(0.86)	
North West	0.03	2.926	−0.67**	−48.63	−0.58*	−43.89
	(0.12)		(−2.62)		(−2.52)	
Gauteng	−0.19	−17.31	−0.25	−22.24	0.18	20.10
	(−1.02)		(−1.34)		(1.03)	
Mpumlanaga	0.14	14.51	−0.78**	−54.09	0.22	24.27
	(0.60)		(−3.10)		(1.05)	
Limpopo	−0.28	−24.62	−0.87***	−58.19	−0.32	−27.25
	(−1.34)		(−3.71)		(−1.38)	
cut1						
Constant	2.37***		2.97***		−0.42	
cut2						
Constant	3.50***		5.06***		2.54***	
cut3						
Constant	7.17***		6.92***		4.00***	
N	**3123**		**3123**		**3123**	

Statistical significance is denoted at (*)10%, (**)5% , and (***)1%

may impact on the use of higher levels of financial services, which is also an important finding emerging from our study.

Finally, in line with the second aim of decomposing the coefficient of gender to identify the relative contributions by economic characteristics and behavioural differences, we employed the non-linear version of the Blinder-Oaxaca decomposition technique (Sinning et al., 2008) in Tables A5.1, A5.2, A5.3, A5.4, A5.5 and A5.6 for Ghana and South Africa. The results show that over 70% of the female coefficient in the general account and unobserved characteristics explain insurance models of gender. With investment services, about 47% of the gender differences are explained by unobserved characteristics. This implies that the variation across gender in terms of financial use is explained by factors other than what has been included in the model. Thus, gendered norms, including habitus that generate risk taking, aggressive masculine traits sustains and reproduces practices in the financial services sector while feminine traits that foster social relations through caring, altruistic and emotional investments are less privileged. The findings suggest that gender is key to financial behaviour, as financial fields value and reward actions that invest in high risk financial services compared to investment in basic financial and insurance services.

We argue that it is important to situate gendered behaviour patterns through a deconstruction of the nature of risk vis-à-vis understanding social and gendered relations both within the household and beyond and differences in geographical contexts. Recent scholarship on the economic crises point to how 'coordinates of the debates' have 'externalized and objectified the economy as a separate domain from society' (Dinerstein et al., 2014). Thus, non-financial approaches for tackling gender differences are required to tackle gender biases in policy approaches. This moves away from a biological basis for observed differences in financial risk taking and deconstructs observed gender differences within the wider framework of social relations which situates women's decision to prioritise investments and the use of financial services that cover social risks than mere participation in higher economic returns.

5.5 Conclusions

The observations on gender gaps in financial behaviour in developing contexts motivate this chapter. Debates on feminisation of finance refer to the challenges in the existing social relations of finance, income gaps and the growing commodification of women's capacities through increasing economic opportunities for women. The question of the 'sub-altern' financial subject, mainly women and racial minorities, is gaining more prominence, and made more stark through the effects of the global financial crises, thus inviting a deconstruction of financial behaviour.

To understand more, we have used FinScope data to examine gender differentials in their access to and use of a range of financial products between Ghana and South Africa. Our findings on financial behaviour between men and women in Ghana and South Africa show that gender is central to the field of financial services. Firstly, the main observation is that women in Ghana are less likely to use general financial and investment products, as compared to South Africa. Secondly, there is some level of gendered risk taking behaviour as men in both countries use advanced investment products, and to some extent the lower use of formal financial services by women can be explained by gender gaps in social dimensions such as lower levels of income, education, regional variation, race and age. This indicate that efforts to bridge the gender income gap need to be addressed to promote use of financial service by all groups of women. Furthermore, the study offers a platform for future research in addressing the issue of endogeneity due to the bi-causal relationship between income and the use of financial services and on the contribution of the individual variables to the observed differences in the coefficient of gender.

Surveying the literature on gendered access to different types of formal finance, this chapter contributes to investigating the presence of gender differentials in using higher order financial services which supports the findings that men use advanced investment products. Financial fields also value and reward behaviour that invest in high risk financial services, further precipitating gendered use of financial services in Ghana. Theories of intra-household economic behaviour point to the inequalities in bargaining power and decision-making where actors operate on altruistic behaviour rather than guided by motives of self-interest or economic rationality. This difference is observed not only in the household, but beyond in the market, community and the state (See Agarwal, 1997). Yet, while previous research points to how gender plays a key role in economic and financial risk aversion behaviours, this chapter shows how financial behaviour can also be transformed through stimuli of social change such as income or education.

In particular, this study pays close attention to how women and men deploy risk while choosing financial services, thus the embodied behaviours derived from the internalisation of the social structure (Bourdieu, 1977, 1984) may change across agents from different social groups and different social outcomes through processes of social change such as education, age and race. As Marlow and Swail (2014) point out 'an association between gender, women and risk avoidance informs a valorization process that axiomatically presumes women are overly cautious and disadvantageous' pointing to women as a 'limited' actor in the wider economic and financial field. Thus, in a policy context, current global discourses on financial inclusion should reflect concerns raised by gender discourses about biases in policies and practices that affect women's meaningful take up and use of financial services. This includes non-financial approaches that deconstruct complex social relations and tackle the gendered material inequalities, primarily income, that determines financial behaviours in developing regions.

Although the sample used in this chapter is large, one limitation of this study can be that we may not generalise the empirical findings to other segments of the population that were excluded from this study due to lack of data. Thus, the results presented in this chapter provide a starting point for conducting future research into how social divisions such as race, ethnicity and class intersect with gendered economic behaviour in developing economies.

5.6 Appendices

See Tables A5.1, A5.2, A5.3, A5.4, A5.5 and A5.6.

Table A5.1 Decomposition of general account and services model—Ghana

	Coef	*Percentage*	*Std. Err*	*Z*
Explained	−0.06	32.36	0.02	−3.14
Unexplained	−0.12	67.64	0.04	−2.87
Total	−0.18	100.00	0.04	−4.65

Note Bootstrap standard errors; 100 repetitions

Table A5.2 Decomposition of insurance model—Ghana

	Coef	*Percentage*	*Std. Err*	*Z*
Explained	−0.12	23.25	0.03	−4.35
Unexplained	0.11	76.75	0.06	1.70
Total	−0.21	100.00	0.09	−2.67

Note Bootstrap standard errors; 100 repetitions

Table A5.3 Decomposition of investment model—Ghana

	Coef	*Percentage*	*Std. Err*	*Z*
Explained	−0.06	53.42	0.02	−2.43
Unexplained	−0.19	46.58	0.09	−2.09
Total	−0.24	100.00	0.09	−2.67

Note Bootstrap standard errors; 100 repetitions

Table A5.4 Decomposition of general account and services—South Africa

	Coef	*Percentage*	*Std. Err*	*Z*
Explained	−0.11	3451.13	0.03	−3.96
Unexplained	0.11	−3351.13	0.06	1.75
Total	0.00	100.00	0.07	−0.05

Note Bootstrap standard errors; 100 repetitions

Table A5.5 Decomposition of insurance model—South Africa

	Coef	*Percentage*	*Std. Err*	*Z*
Explained	−0.11	1086.52	0.02	−5.00
Unexplained	0.10	−986.52	0.06	1.68
Total	−0.01	100.00	0.07	−0.15

Note Bootstrap standard errors; 100 repetitions

Table A5.6 Decomposition of investment model—South Africa

	Coef	*Percentage*	*Std. Err*	*Z*
Explained	−0.01	10.91	0.02	−0.63
Unexplained	−0.12	89.09	0.15	−0.78
Total	−0.13	1.00	0.15	−0.90

Note Bootstrap standard errors; 100 repetitions

References

Agarwal, B. (1997). Bargaining and gender relations: Within and beyond the household. *Feminist Economics, 3*(1), 1–51.

Albertyn, C. (2011). Law, gender and inequality in South Africa. *Oxford Development Studies, 39*(02), 139–162.

Allen, F., Carletti, E., Cull, R., Qian, J. Q., Senbet, L., & Valenzuela, P. (2014). The African financial development and financial inclusion gaps. *Journal of African Economies, 23*, 614–642.

Angel-Urdinola, D., & Wodon, Q. (2010). Income generation and intra-household decision making: A gender analysis for Nigeria, *MPRA Paper* No. 27738

Annim, S., Arun, T., & Kostov, P. (2012). Effect of perceptions and behaviour on access to and use of financial service: Evidence from South Africa, *IZA Discussion Paper Series* No. 7042.

Arnold, J., & Gammage, S. (2019). Gender and financial inclusion: The critical role for holistic programming. *Development in Practice, 29*(8), 965–973.

Arun, S., Annim, S. & Arun, T. (2016). 'Even' after access to financial services? Ricocheting gender equations, *IZA Discussion Paper Series* No. 10099.

Arun, T., & Kamath, R. (2015). Financial inclusion: Policies and practices. *IIMB Management Review, 27*(4), 267–287.

Ashraf, N. (2009). Spousal control and intra-household decision making: An experimental study in the Philippines. *American Economic Review, 99*(4), 1245–1277.

Assassi, L. (2009). *The gendering of global finance.* Springer.

Asuming, P. A., Osei-Agyei, L. G., & Mohammed, J. I. (2019). Financial inclusion in Sub-Saharan Africa: Recent trends and determinants. *Journal of African Business, 20*(1), 112–134.

Aterido R., Beck T., & Iacovone L. (2011). Gender and finance in Sub-Saharan Africa: Are women disadvantaged? *World Bank Policy Research Working Paper* 5571.

Barbarin, O. A., & Richter, L. M. (2013). *Mandela's children: Growing up in post-apartheid South Africa.* Routledge.

Barslund, M., & Tarp, F. (2008). Formal and informal rural credit in four provinces of Vietnam. *The Journal of Development Studies, 44*(4), 485–503.

Beck, T., Senbet, L., & Simbanegavi, W. (2015). Financial inclusion and innovation in Africa: An overview. *Journal of African Economies, 24*(suppl 1), i3–i11.

Bliss, R. T., & Potter, M. E. (2002). Mutual fund managers: Does gender matter? *The Journal of Business and Economic Studies, 8*(1).

Bourdieu, P. (1977). *Outline of a theory of practice.* Cambridge University Press.

Bourdieu, P. (1984). *Distinction: A social critique of the judgement of taste.* Harvard University Press.

Bourdieu, P., & Wacquant, L. J. (1992). *An invitation to reflexive sociology.* University of Chicago Press.

Burchi, F., & Vicari, S. (2014). To be or not to be a member of a primary co-operative in Brazil: Any difference in household decision-making and gender equality? *Oxford Development Studies, 42*(3), 343–364.

Byrakatar, N., & Fofack, H. (2018). A model for gender analysis with informal production and financial sectors. *Journal of African Development*, *20*(2). Penn State University Press.

Calkin, S. (2015). Feminism, interrupted? Gender and development in the era of 'Smart Economics.' *Progress in Development Studies, 15*(4), 295–307.

Chant, S., & Sweetman, C. (2012). Fixing women or fixing the world? 'Smart economics', efficiency approaches, and gender equality in development. *Gender & Development, 20*(3), 517–529.

Charness, G., & Geezy, U. (2012). Strong evidence for gender differences in risk taking. *Journal of Economic Behaviour and Organization, 83*, 50–58.

Chikalipah, S. (2017). What determines financial inclusion in Sub-Saharan Africa? *African Journal of Economic and Management Studies, 8*, 8–18.

Deghaye, N., McKenzie, T., & Chirawu, P. (2014). Inequality in South Africa: A two part document on the current understanding and dimensions of inequality in health, gender and livelihoods.

Dinerstein, A.C., Schwartz, G., & Taylor, G. (2014). Sociological imagination as social critique: Interrogating the 'global economic crisis'.

Doss, C. (2013). Intrahousehold bargaining and resource allocation in developing countries. *World Bank Research Observer, 28*(1), 52–78.

Doyle, J. A., & Paludi, M. A. (1991). *Sex and gender: The human experience.* William C. Brown.

Eckel, C. C., & Grossman, P. J. (2002). Sex differences and statistical stereotyping in attitudes toward financial risk. *Evolution and Human Behaviour, 23*(4), 281–295.

Evans, O. (2018). Connecting the poor: The internet, mobile phones and financial inclusion in Africa. *Digital Policy, Regulation and Governance, 20*, 568–581.

Finmark Trust. (2011). *FinScope South Africa Small Business Survey 2010.* Finmark Trust.

Folbre, N. (1986). Hearts and spades: Paradigms of household economics. *World Development, 14*, 245–255.

Global Findex Database. (2014, 2015). Measuring financial inclusion around the world. *World Bank Policy Research Working Paper* 7255.

Heintz, J. (2005). Employment, poverty, and gender in Ghana. *PERI Working Papers*, 66.

Katz, E. (1997). The intra-household economics of voice and exit. *Feminist Economics, 3*(3), 25–46.

Kostov, P., Arun, T., & Annim, S. (2011). Determinants of access to finance: An investigation into the Mzansi intervention. *European Journal of Development Research, 24*(3), 397–412.

Lewis, R., Villasenor, J., & West, D. M. (2016). *Bridging the financial inclusion gender gap*. The Brookings.

Maclean, K. (2010). Capitalizing on women's social capital? Women-targeted microfinance in Bolivia. *Development and Change, 41*(3), 495–515.

Marlow, S., & Swail, J. (2014). Gender, risk and finance: Why can't a woman be more like a man? *Entrepreneurship & Regional Development, 26*(1–2), 80–96.

Marlow, S., & Swail, J. (2015). Re-evaluating business exit from a gendered perspective. In DeTienne D., Wennberg K. (Eds.), *Research Handbook of Entrepreneurial Exit* (pp. 42–62). Edward Elgar.

McElroy, M., & Horney, M. J. (1981). Nash-bargained household decisions: Towards a generalization of the theory of demand. *International Economic Review, 22*(2), 333–349.

Meier-Pesti, K., & Penz, E. (2008). Sex or gender? Expanding the sex-based view by introducing masculinity and femininity as predictors of financial risk taking. *Journal of Economic Psychology, 29*(2), 180–196.

Morsy, H. (2020). Access to finance: Why aren't women leaning in? Women are self-selecting out of the African credit market. *Finance and Development, 57*(1).

Morsy, H., & Youssef, H. (2017). Access to finance—Mind the gender gap. *EBRD Working Paper* 202, European Bank for Reconstruction and Development, London.

Olsen, R. A., & Cox, C. M. (2001). The influence of gender on the perception and response to investment risk: The case of professional investors. *Journal of Psychology and Financial Markets, 2*(1), 29–36.

Sen, A. (1990). Gender and cooperative conflicts. In I. Tinker (Ed.), *Persistent inequalities: Women and world development*. Oxford University Press.

Sinning, M., Hahn, M., & Bauer, T. K. (2008). The Blinder-Oaxaca decomposition for nonlinear regression models. *The Stata Journal, 8*(4), 480–492.

UNDP. (2014). The impacts of social AND economic inequality on economic development in South Africa.

Van Hulten, A. (2012). Women's access to SME finance in Australia. *International Journal of Gender and Entrepreneurship, 4*(3), 266–288.

Zins, A., & Weill, L. (2016). The determinants of financial inclusion in Africa. *Review of Development Finance, 6*, 46–57.

CHAPTER 6

Bank Competition and Financial Sector Stability in Ghana

William Godfred Cantah, Kwabena Nkansah Darfor, Jacob Nunoo, Benedict Afful Jr., and Emmanuel Agyapong Wiafe

6.1 Introduction

Prior to 1986 Ghana's banking industry was characterised by direct controls by the government and was dominated by activities of state-owned banking institutions with very little or no competition. All these changed gradually when the Financial Sector Adjustment Programme

W. G. Cantah (✉)
Department of Data Science and Economic Policy, School of Economics, University of Cape Coast, Cape Coast, Ghana
e-mail: william.cantah@ucc.edu.gh

K. N. Darfor · J. Nunoo
Department of Applied Economics, School of Economics, University of Cape Coast, Cape Coast, Ghana
e-mail: kdarfor@ucc.edu.gh

J. Nunoo
e-mail: jacob.nunoo@ucc.edu.gh

J. A. Peprah et al. (eds.), *Financial Sector Development in Ghana*, Palgrave Macmillan Studies in Banking and Financial Institutions,
https://doi.org/10.1007/978-3-031-09345-6_6

(FINSAP) which began in 1986 was implemented. The FINSAP implementation removed direct state controls in the banking industry. In addition, several state-owned banking institutions were privatised in order to increase the competitiveness and efficiency of the banking industry. To further enhance competition and stability of Ghana's banking industry, there was the implementation of the Financial Sector Strategic Plan (FINSSP) in 2003. FINSSP led to a more liberalised financial sector with universal banking licences also granted to several financial institutions. The number of banking institutions in the country increased from 10 (dominated by state-owned banks) in 1989, to 29 (dominated by privately owned banks including foreign-owned ones) by April 2013.

In addition to the efforts aimed at enhancing competition and efficiency in the banking industry, several efforts were also made to sustain the stability of the industry to propel the growth and development of the economy. Laws such as the Banking Act 2004 (Act 673), Credit Reporting Act, 2006 (Act 726), Banking Amendment Act 2007 (Act 738), Borrowers and Lenders Act 2008 (Act 773) and the Ghana Deposit Protection Act 2016 (Act 931) were all enacted to ensure the continued stability of the financial sector. Furthermore, the Bank of Ghana from time to time increased the minimum capital requirement for banking institutions to ensure the continuous stability of the sector. The minimum capital requirement was increased from GH¢7million (US$1,211,072.66) in 2007 to GH¢60million (US$ 10,380,622.80) in 2009, then to GH¢120million (US$20,761,245.60) in 2017 and finally to GH¢400million by close of 2018.

Despite all these efforts to ensure its stability, Ghana's banking sector remained fragile. In August 2017, two commercial banks (UT Bank and Capital Bank) were on the verge of collapsing, hence, to salvage the situation, GCB bank Limited was made to take over the good assets

B. Afful Jr.
Department of Applied Economics, School of Economics, University of Cape Coast, Cape Coast, Ghana
e-mail: bafful@ucc.edu.gh

E. A. Wiafe
Department of Economics, Ghana Institute of Management and Public Administration, Accra, Ghana
e-mail: ewiafe@gimpa.edu.gh

of these two financial institutions to ensure the safety of depositors' funds. In 2018, five other banks (Unibank, Construction Bank, Sovereign Bank, Beige Bank and Royal bank) were merged into the Consolidated Bank Ghana limited. The takeover of two banks by GCB coupled with the consolidation of five additional banks within the space of two years increased the concerns of stakeholders in relation to the stability of the industry and how the reduction in the number of banks would affect overall competitiveness in the sector.

Implementation of the FINSAP and the FINSSP I & II were expected to increase competition (through liberalisation) and stability of the financial sector. However, since the implementation of these reforms, very little has been done to empirically verify the extent to which a rise in the competitiveness of the banking sector has added to its stability. This has become particularly necessary following the recent development in the industry as explained above. Thus, there is the need for us to empirically understand how banking industry competitiveness tends to affect its stability in particular and the entire financial landscape.

The attainment of financial stability and the role of competition in making that possible have dominated research over the last two decades (Kasman & Kasman, 2015). Competition in any sector of the economy mostly enhances innovation, efficiency and greater product quality. However, the effect of banking sector competition on the stability of the financial sector is still controversial. There are two (competition-fragility & competition-stability) main competing views in the banking institutions competition literature. Competition-fragility is based on the argument that higher competition in the banking sector tends to reduce market power and profit margins, as such banks would have greater incentive to take higher risk to boost profitability (Keeley, 1990; Marcus, 1984). As banks take on more risk, asset quality and efficiency are jeopardised. These high-risk businesses may also result in instability in the banking sector. Therefore, in this respect, bank competition introduces high probability of instability in the banking industry. On the other hand, the competition-stability view as proposed by De Nicoló et al. (2006) argues that higher level competitiveness results in lower interest rates in the market and reduction in adverse selection and moral hazard issues among borrowers. This they maintain would cause a reduction in loan default and enhance stability. Another school of thought believes that it is plausible for these two conflicting views to be put together to generate a U-Shaped relationship between competition and stability (Martinez-Miera & Repullo, 2010).

Over the years, the competitiveness and stability relationship has dominated research in developed as well as developing countries (Amidu & Wolfe, 2013; Beck, 2008; Clark et al., 2004; De-Ramon et al., 2018; Gudmundsson et al., 2013; Kasman & Kasman, 2015; Oduor et al., 2017). Empirical evidence from existing research is mixed, with some studies supporting the competition-fragility and others supporting the competition-stability view. Some studies have identified the structure of the banking industry (level of concentration) as key determinant of stability (Fernández & Garza-García, 2015; Fu et al., 2014). Given that results from existing studies are mixed and the fact that the structure of a country's financial system is key in determining the extent to which competition affects stability, the present study seeks to add to the literature by analysing the extent to which competition affects the stability of Ghana's banking industry.

The recent banking crisis offers an opportunity to examine how various interventions to liberalise Ghana's financial space have impacted on stability. Existing studies on competition either focus on competition and profitability (Ahiabor, 2013) or competition and efficiency using data envelopment analysis (Adjei-Frimpong et al., 2016; Alhassan & Ohene-Asare, 2016). The current study adds to the existing literature by providing empirical evidence from Ghana on competition, stability/fragility subject. In doing so, the study examines whether competition-fragility or competition-stability viewpoints in the literature hold in the case of Ghana. The study further employs a panel threshold regression to examine whether the extent of concentration in the banking sector underpins the relationship between competitiveness and stability. This is novel in the bank competition-stability literature. In addition to this, the use of the panel threshold regression would help to identify the specific level of competitiveness in the banking sector that would enhance stability in the sector. Thus, making the findings of this study more practical to policymakers by providing a range of concentration that policy should target to ensure stability in the financial space in Ghana. In other words, we are able to point out the critical levels of concentration in moderating the competition-stability nexus in Ghana.

6.2 Review of Related Literature

This section discusses the theoretical and empirical studies that underpin the work. The theoretical literature hinges on the competition-fragility, competition-stability and the U-shaped link between competition and stability. Competition-fragility posits that competitiveness results in less market power and profits, while the competition-stability view argues that competition enhances market power. In another vein, the predicted U-shaped link between competitiveness and stability combines the two contrary views. The view of fragility is supported by Petersen and Rajan (1995), who stress that enhanced competitiveness lowers the supply of credit, since the likelihood of banks granting credit to non-established customers is very low. In their view, decline in the power of the market reduces the motivation for banks to stick with their existing customers which could stifle chances of increasing avenues for shared surpluses. Furthermore, Saurina Salas et al. (2007) argue that risky loan portfolios are more prevalent in competitive banking environments. Hellmann et al. (2000) indicate elimination of ceilings on interest rate engendering price competitiveness, reduced value of franchise and fuel clients' moral hazard behaviour. Again, in a study by Keeley (1990) higher banking competitiveness together with deregulation in the United States of America in the 1990s were found to result in reduction in monopoly rents and as well as contributing to fragility of banks.

In another vein, competition-stability is reinforced by studies such as Love and Martinez-Peria (2012) and Hainz et al. (2013) who argue a positive link between competition, liquidity creation and financial stability. The U-shaped relationship between competition and stability is corroborated by Martinez-Miera and Repullo (2010) who argue that competition leads to a fall in market power and profit, attains a minimum and thereafter rises (rise in market power and profit).

On the empirical level, there are several studies in this regard. Horvath et al. (2016) using dynamic GMM panel estimations with data from Czech Republic found that increased competitiveness leads to a fall in liquidity creation. They explained further that rising competitiveness of banks impacts on the financial fragility of banks, which causes banks to lower their lending as well as deposit activities and by extension to decreasing their (banks) market power and profit margins. On the other hand, Fernández and Garza-García (2015) examined the hypotheses on competition-fragility and competition-stability in the banking sector of

Mexican between the period 2001 and 2008. The study proxied financial stability and bank portfolio risks by the Z-index and ratio of non-performing loans to total loans respectively, while the level of competition was proxied by the Lerner index. The findings of this study confirmed the existence of both views for Mexico. However, they were quick to add that, benefits derived from increased competitiveness on the stability of the entire financial system far outweighed the rise in overall fragility.

Berger et al. (2017) found evidence for both hypotheses using data covering 23 developed nations. They found arguments that supported the competition-fragility view, which suggest that the more market power increased the risk exposure of banks also reduced. Furthermore, Liu et al. (2013) investigated the competitiveness in 11 EU countries between the period 2000 and 2008 to analyse the competition-stability relationship in the banking sector. The Lerner index and the Z-index were used as proxies for bank competition and bank stability respectively by the authors. The result revealed that within the European banking sector a non-linear relationship exists between competition and stability. Distinctively, they found that in highly concentrated markets, there were risk-shifting effects, where improved bank competitiveness reduced net interest margins and led to a rise in bank stability. They also established that in the face of heightened competition, marginal effects exist, leading to reduced loan interest payments and non-performing loans.

In addition, in their cross-country study, Schaeck et al. (2009) applied Logit regression and duration analysis in estimating competitiveness which was captured by the Rosse-Panzar H-statistic. They found that there was a lower likelihood of bank failure in competitive banking sectors, than in monopolistic banking systems. Ariss (2010) also considered the extent to which the power of the market influenced efficiency and stability of the financial sector for a group of emerging economies. Applying different specifications of the Lerner index to capture competition and Z-index to capture stability of the financial sector, the study findings showed that a rise in the power of the market leads to increased stability of banks for emerging countries, even though it is associated with loss in cost efficiency.

In Ghana, Ahiabor (2013) assessed bank competition and profitability from 2007 to 2012 using the System Generalized Method of Moment (GMM) and found evidence that the banking sector is quite competitive with a restrained level of persistence of profit. He observed that credit risk (ratio of non-performing loans) was the only bank-specific determinant influencing bank profitability significantly. This study further revealed that annual GDP growth rate and inflation were the two macroeconomic factors influencing the profitability of banks in Ghana. In their study, Buchs and Mathisen (2005) measured the level of competition and efficiency of banks in Ghana from 1998 to 2003. Employing the Panzar-Rosse model as a measure of bank competition, they found a Panzar-Rosse value of 0.56 indicating that the Ghanaian Banking sector was non-competitive. They also established that, the main determinants of bank profitability in Ghana are high interest rate, bank size, expense ratio and inflation. Pino and Araya (2013) also looked at the effect of heterogeneity in the power of the market on the link involving competition and stability for the banking sector of Chile. Their results indicated that the link involving risk-taking and competition is significantly influenced by heterogeneity. Using data from Turkey spanning 2002–2012 and the Boone indicator as well as the efficiency-adjusted Lerner index as proxies for competition and non-performing loans (NPL) ratio and Z-scores as proxies for banking stability, Kasman and Kasman (2015) found that competition was negatively related to the NPL ratio but positively related to the Z-score. The results from their study further indicated that greater concentration has a direct effect on non-performing loans and inversely impact on the Z-score. They introduced a quadratic term for competition to test for a possible non-linear link between competition and stability. They introduced a quadratic term for competition to test for a possible non-linear link between competition and stability. The finding indicated that the estimated parameter of the quadratic term is negative for the non-performing loan model and positive for the Z-score model. Overall, their findings lend credence to the competition-fragility hypotheses. Schaeck et al. (2009) also provided empirical evidence for the trade-off between competition and banks' risk-taking behaviour using data from 45 countries. Their findings suggest that when operating in a more competitive environment and competitive banking markets, banks have the tendency to hold higher capital reserves and are less likely to experience systemic crisis. The literature reviewed thus far reveals mixed results regarding the relationship between competition and concentration can coexist and influence stability through different channels.

So far, the study that comes close to the current one is Kasman and Kasman (2015) which tested for the existence of a non-linear link between competition and stability. However, their work did not emphasise the influence of concentration on the nexus between competition and stability. Currently, we haven't sighted any study that uses panel threshold model to examine the question at hand. However, financial time series data may exhibit higher seasonality, breaks and jumping effect which could be captured well using a threshold regression model. In furtherance to the above, studies by Ahiabor (2013) and Buchs and Mathias (2005) had their occupations solely on efficiency and profitability of banks in Ghana. Though Buchs and Mathias (2005) touched on competition, they basically examined the levels of competitiveness in the banking space of Ghana. Given the recent banking sector bubbles in Ghana and the financial sector reforms over the years, it is prudent to examine empirically the role of competition on stability. Since the cardinal goal of the financial sector reforms was to encourage efficiency, financial sector stability and competition through liberalisation of the financial sector of Ghana, this study is crucial in answering the question regarding how these reforms have contributed to financial stability in Ghana.

6.2.1 *Stylized Facts and Data*

The banking sector in Ghana has gone through various phases over the years. Bank concentration, competition, market power, liquidity and size were the variables used for the study. Bank assets in Ghana have experienced sustained annual growth for three consecutive years with the total banking size measured by asset growing on the average of 13.8% in 2019 (BOG, 2019) after the banking sector crisis. The total assets of banks for the period 2017 to 2019 were 92.2% made up of domestic investment as against 7.8% foreign assets (BOG, 2018, 2019). This shows a strong domestic dependency of the banking sector. The implication is that an economic recession may have dire consequences on the asset build-up of the banks and lead to capital inadequacy. Bank liquidity is crucial for the sustainability of financial institutions and their ability to meet future financial obligations when they are due. Over time, liquidity risk may culminate into market risk. Hence, improvement in liquidity of banks influences bank stability. Data available shows some level of gains in liquidity of banks since the 2018 banking sector reforms. The liquidity of bank reserves to banking assets has been on an upward trend over the past decades with the exception of 2002 where it was at its lowest point. The highest ratio recorded was 38.6 according to IMF data. Data from the Bank of Ghana show steady improvement in the liquidity of banks from 54.2 to 62.5 when liquid assets are compared with total assets in their broad base (Bank of Ghana, 2020). Similar indicators show same trends. For instance, the liquid asset to total deposit was around 99.2%.

Events such as the COVID-19 pandemic that have led to collapse of firms and contributing to unemployment may have a potential risk on bank assets especially deposit contribution. Thus, banks' operating assets may be affected by the COVID-19 pandemic in an adverse manner. Nevertheless, other indicators show a more resilient banking sector despite the possibility of banks rebounding into crisis in the post-COVID-19 period.

The banking industry in Ghana experiences increased cases of non-performing loans which is a fundamental problem for banks. The implication of such rising levels of non-performing loans for banks in Ghana is instability and subsequent liquidation and insolvency of the banks. Trends show a banking industry with average non-performing loans of about 15%. Compared to 2009, 2017 saw the highest level of non-performing loans in a decade (21.59%). The sharp rise in the level of NPL in 2017 could possibly be attributed to the inability of banks that were winding-up their activities to call in overdue loans. This subsequently dropped to 18.19% in 2018. In 2019, the non-performing loans of banks further declined to 13.9% (BOG, 2020). This was achieved at the back of sound financial and risk management practices.

The banking sector Z-score serves as indication for relating the bank capital to the variability in its returns. This helps to ascertain the amount of variation in returns that can be dealt with by capital without the bank suffering insolvent. From Fig. 6.1, on the average, between 2011 and 2013 the bank Z-score increased moderately from little above 16.1 to 17.1% indicating moderate improvement in banking sector stability between these periods.

As shown in Fig. 6.1, between 2013 and 2014 there was a sharp rise in banking sector stability from 17.1 to about 22.1% which is the highest within the period under study. From the period 2015 to 2018, the banking sector suffered from instability which could reduce trust and serve as an indication of insolvency especially in 2016.

NPLs is measured as the amount of non-performing loans divided by total amount of outstanding loans the banks hold. This shows that a rise in the ratio signifies poor performance of loans. From Fig. 6.2, it can be observed that between 2011 and 2014 the non-performing loans were relatively lower compared to the period 2015–2018 where the non-performing loans were relatively higher. From Fig. 6.2, it is evident that,

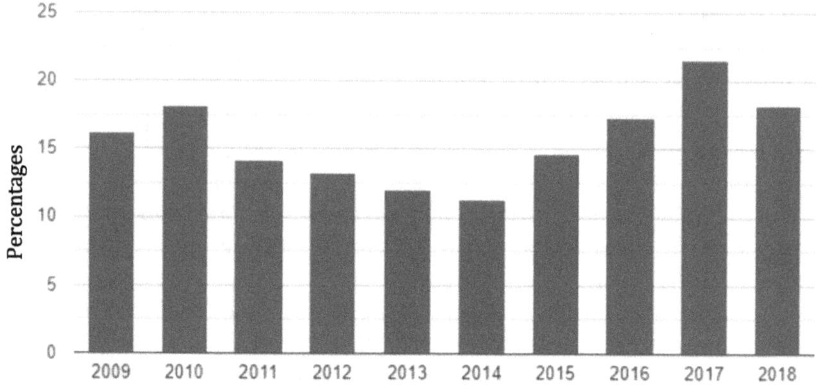

Fig. 6.1 Non-performing loans as percentage of total assets (*Source* World Financial Indicators)

on the average, in periods of low non-performing loans there is high level of bank stability measured using Z-score and during periods of high non-performing loans the reverse is true.

Market power measures the firm's ability to increase its price beyond the marginal cost or the level that might exist under perfect competition. This normally results from reductions in output in order to increase

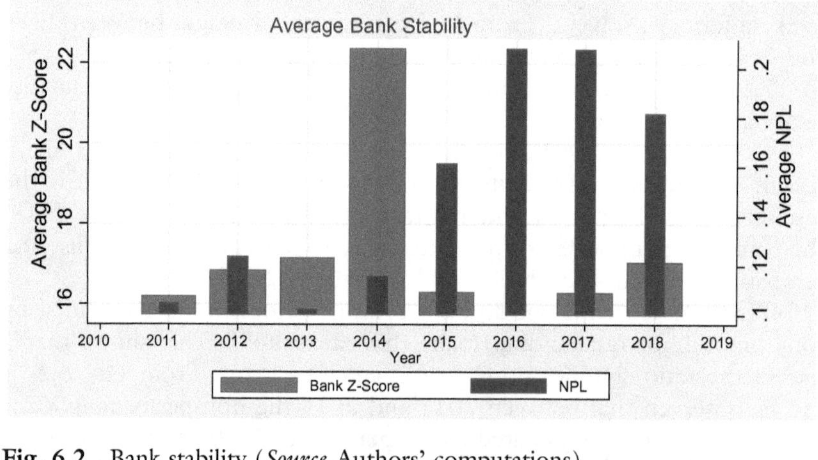

Fig. 6.2 Bank stability (*Source* Authors' computations)

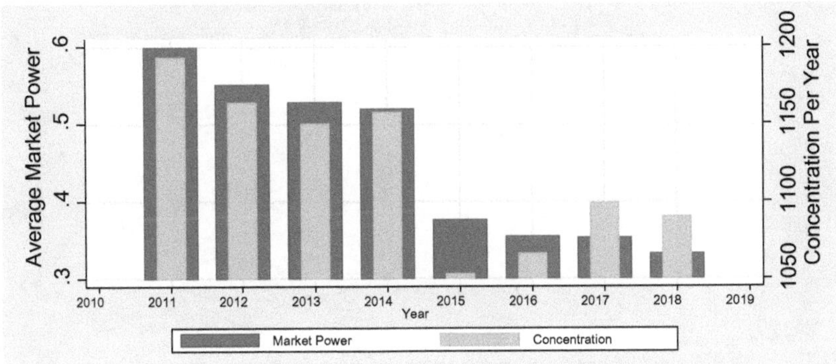

Fig. 6.3 Nature of competition (*Source* Authors' computation)

price. While market concentration measures the number of firms and their respective shares in an industry (total production). From Fig. 6.3, it can be observed that, between 2011 and 2014, market power and concentration were relatively high even though they taper off and subsequently fluctuate beyond 2014. On average market power reduced from 6.0 to 3.4 from the period 2011 to 2018 but market concentration reduced from a little below 1200 in 2011 to about 1050 in 2015 and started rising in 2016. On average we can say that market power in the banking industry has been reduced over the period under study.

From Fig. 6.4, it can be observed that bank stability has been on the decline over the period of study in Ghana, from as high as 22.1 in 2011 to as low as 16.5 in 2018. While market power exhibits a bell shape distribution rising from 16.2 in 2011 and peaking at 2014 with a value above 22.0 and falling below 16.0 in 2016.

6.3 Methodology

To analyse the link between competition and bank stability in Ghana, the study estimated the fixed and random effects model. The study also employed a panel threshold regression to determine the extent to which the relationship between competition and stability is moderated by concentration in the banking sector. The study employed the regression

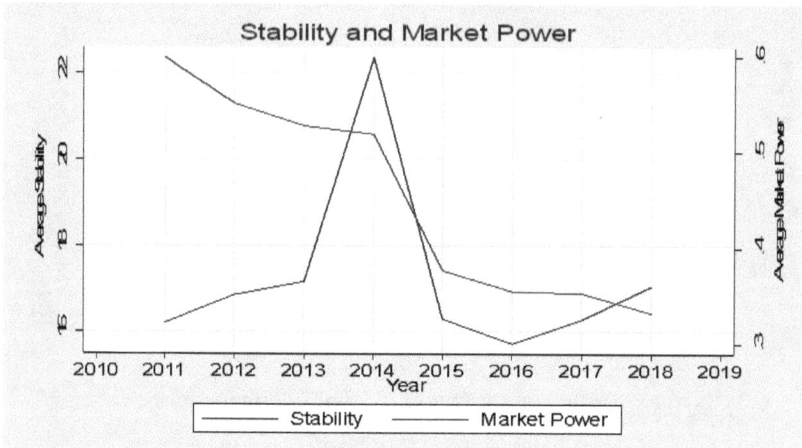

Fig. 6.4 Stability and market power (*Source* Authors' computation)

model captured in Eq. 6.1 as follows:

$$\text{Stability}_{it} = \beta_0 + \beta_1\text{Competition}_{it} + \beta_2\text{Size}_{it} + \beta_3\text{Liquidity}_{it} + \beta_4\text{Concentration}_{it} + \varepsilon_{it} \quad (6.1)$$

The subscripts i and t represents bank and year respectively. We measured stability using the bank Z-score and the percentage of NPLs, while Adjusted Lerner Index was used as a proxy for competition. The choice of Adjusted Lerner was basically informed by the argument of Koetter et al. (2012) who indicated that the conventional Lerner Index implicitly assumes the existence of profit efficiency and cost efficiency and fails to take into account the possibility that banks may not exploit pricing opportunities resulting from market power. We measured bank size using the natural log of total assets of each bank, while Liquidity was measured by dividing total loans of the bank by its total assets. Bank concentration was measured by dividing the total assets of a bank in a particular year by the total assets of the entire banking industry for that same year $\left(\text{i.e. Con} = \left(\frac{TA_t}{ITA_t}\right) \times 100\right)$, where, TA is the bank's asset at time t and ITA is the industry's asset as time t.

Since the study sought to identify the influence of concentration in moderating the link involving competition and stability, the study employed panel threshold regression developed by Wang (2015). The threshold regression was specified as follows:

$$\begin{aligned}\text{Stability}_{it} &= \beta_0 + X_{it}(q_{it} < \gamma)\beta_1 + X_{it}(q_{it} \geq \gamma)\beta_2\text{Size}_{it} \\ &\quad + \beta_4\text{Liquidity}_{it} + \beta_5\text{Concentration}_{it} + u_i + \varepsilon_{it} \qquad (6.2)\end{aligned}$$

where q_{it} is the threshold variable (concentration), and γ is the threshold parameter that divides the equation into two regimes with coefficients β_1 and β_2 (which captures the different effects of competition on stability). u_i is the parameter for the individual effects while ε_{it} is the error term.

6.3.1 *Measure of Competition*

Competition (or Market Power) was captured by the Adjusted Lerner Index. The Adjusted Lerner Index lies between the values 0 (perfect competition) and 1 (pure monopoly). The time-variant Adjusted Lerner index (ALI) on a bank level is calculated as follows:

$$\text{ALI}_{it} = \frac{(\pi_{it} + TC_{it}) - (MC_{it} \times Q_{it})}{\pi_{it} + TC_{it}}$$

where π_{it} represents the profit of bank i at time t, TC_{it} is the total cost of bank i at time t, MC_{it} is the marginal cost of bank i at time t and Q_{it} is the total output of bank i at time t. We derive the marginal cost from the translog total cost function by differentiating the total cost function with respect to output.

6.3.2 *Bank Stability Measure*

As indicated earlier, the study employed two different measures of stability: percentage of non-performing loans and the Z-score. The Z-score is a commonly used measure of bank stability and can be seen as the number of standard deviations by which returns would have to fall from the mean to deplete the equity capital (Boyd & Runkle, 1993). Thus, it determines the degree to which returns on a bank's assets deviates from the level of insolvency. The Z-score is computed as:

$$\text{BZ}_{it} = \frac{ROA_{it} + \frac{E_{it}}{TA_{it}}}{\sigma_{ROA_{it}}}$$

where ROA_{it} is the return on assets of bank i at time t, E_{it}/TA_{it} represents equity to total assets ratio of bank i at time t and $\sigma_{ROA_{it}}$ denote the standard deviation of return on assets of bank i at time t. When the Z-score is high, it implies that there is a low probability of insolvency.

6.3.3 *Data*

Data employed for the study was obtained from annual financial statements of twelve banks whose total assets together represent about 71% of the total assets in Ghana's banking industry. Thus, our data effectively represent activities in the entire banking industry in Ghana. The data period is from 2011 to 2018. The choice of 2011–2018 is mainly due to data availability.

6.4 Results and Discussions

Table 6.1 presents the summary statistics of the variables used for the study. We have 96 bank-year observations for 12 Ghanaian commercial banks over a period of 8 years (2011–2018). The value of 17.25 is the mean Z-score with a standard deviation of 13.92 which implies marginal variation between the overall and between standard deviations. The variation that exists for the within and between is quite pronounced. There seem to be more variation in the Z-scores of the banks than will be observed over time. There seems to be limited variation in the NPL of banks for both the between and within. Also, it can be observed that there is more variation in market power overtime as compared to the variation in market power among banks. For the liquidity and bank size variables, we observe marginal variations for the between and within.

The regression results are presented in Table 6.2 with bank Z-score as the dependent variable. We estimated four regression models. In the first model, we observe a statistically significant positive link involving competition and financial stability with a coefficient of 11.34. Since Adjusted Lerner Index was our measure of competition, an increase in the value of the index implies high levels of market power. This means that higher levels of market power lead to stability in the banking industry in Ghana. This result supports the competition-fragility arguments of Keeley (1990) and Marcus (1984). This result implies that high levels of competition (low market power) could be inimical to the stability of Ghana's financial sector. The low level of financial development and inclusion in Ghana

Table 6.1 Descriptive statistics

Variable		*Mean*	*Std. Dev*	*Min*	*Max*	*Observations*
Z-Score	Overall	17.25	13.92	0.54	80.97	N = 96
	Between		12.22	5.80	40.87	n = 12
	Within		7.45	5.46	79.40	T = 8
NPL	Overall	0.15	0.10	0.001	0.49	N = 96
	Between		0.07	0.07	0.28	n = 12
	Within		0.08	−0.06	0.36	T = 8
Competition	Overall	0.45	0.15	0.00	0.94	N = 96
	Between		0.06	0.31	0.52	n = 12
	Within		0.14	0.14	0.92	T = 8
Concentration	Overall	8.33	4.93	0.94	20.61	N = 96
	Between		4.89	2.86	18.51	n = 12
	Within		1.47	2.33	12.09	T = 8
Liquidity	Overall	0.47	0.48	0.13	4.94	N = 96
	Between		0.21	0.26	1.09	n = 12
	Within		0.43	−0.18	4.32	T = 8
Size	Overall	21.47	0.81	19.42	23.10	N = 96
	Between		0.57	20.57	22.44	n = 12
	Within		0.60	19.58	22.58	T = 8

Source Authors' computation

relative to developed economies implies that highly competitive banking industry would compel banking institutions to take greater risks to make higher profits which would adversely affect the stability of the industry.

Interestingly, concentration, which was captured as the share of banking assets in the banking industry, was found to have negative and significant impact on stability of the banking sector. High levels of concentration are mostly associated with high cost of credit, extensive diversification, which may not necessarily have direct impact on bank's activities, and also creates lower pricing power and monitoring problems for such banks making them more fragile and leading to instability of the industry as a whole. This finding confirms the concentration-fragility argument of Uhde and Heimeshoff (2009) who argued that moral hazard problems become severe for larger banks. Cetorelli et al. (2007) also maintain that the larger the bank the lower the level of managerial efficiency which in turn affects its level of stability. This finding corroborates the results obtained by Fu et al. (2014) and Calice and Leonida (2018). Model 1 further indicates a statistically significant positive relationship

Table 6.2 Competition and stability estimates

Variables	*(1)* *Bank Z-score*	*(2)* *NPL*	*(3)* *Bank Z-score*	*(4)* *NPL*
Competition	11.34***	−0.195*		
	(4.108)	(0.107)		
Size	2.459*	0.0282	2.725**	0.0301
	(1.495)	(0.0262)	(1.162)	(0.0171)
Liquidity	14.10***	−0.00263	12.61***	−0.0264**
	(1.291)	(0.00984)	(0.790)	(0.0101)
Concentration	−0.778**	−0.0150**	−0.690**	−0.0365***
	(0.313)	(0.00703)	(0.287)	(0.00685)
Competition before threshold			18.23***	−0.223***
			(4.970)	(0.0588)
Competition after threshold			5.552	0.119**
			(3.594)	(0.0499)
Constant	−40.76	−0.240	−45.81	−0.0942
	(34.13)	(0.545)	(26.39)	(0.349)
Threshold value for concentration			5.2799	16.807
Observations	96	96	96	96
R-squared			0.865	0.483
Number of banks	12	12	12	12

Robust standard errors in parentheses; ***$p < 0.01$, **$p < 0.05$, *$p < 0.1$

between bank liquidity and bank Z-score with a parameter value of 14.10. The positive sign for the coefficient of bank size indicates that big banks in Ghana are exposed to lower risks. This finding is at variance with that of Fu et al. (2014) and Kasman and Kasman (2015).

Column 2 of Table 6.2 shows the results of the effect of bank competition and other control variables on stability of the financial sector using NPLs as the measure of financial stability. We specified four models all of which have NPL as the outcome variable. From Model 2, we find a statistically significant inverse relationship between bank competition and non-performing loans implying that increased power of the market results in a decline in NPLs. It can be concluded that increased competitiveness in the banking sector is inimical to the stability of the financial sector. The increased bank competition also feeds into an increase in bank portfolio risks, which confirm the competition-fragility viewpoint. This finding supports the findings of Ruza et al. (2019), Fernández and Garza-García (2015), Fu et al. (2014), Pak and Nurmakhanova (2013) and

Agoraki et al. (2011). However, it contradicts the findings of Yaldiz and Bazzana (2010) and Yeyati and Micco (2007). The result of the Lerner index is consistent with results obtained when concentration is used as a measure of competition. Bank concentration showed a statistically and significant negative link with NPLs indicating that increased bank concentration reduces the exposure of banks to credit risks. This is probably because the concentration of assets provides banks some level of advantage in their loan recovery process since they would be in a position to deploy all the resources within the legal framework of the country to get their defaulting clients to payback their loans. The influence of concentration on credit risk (NPL) contradicts the results obtained in model (1) where concentration was found to have adverse effect on the level of solvency in the industry. This could be attributed to the fact that banks with larger concentration may take on too much risk with the perception that they would be in a better position to recover their investments. However, such increases in risk appetite tend to reduce liquidity available for their operation which eventually reduces their level of solvency.

Furthermore, the results show no statistically significant relationship between liquidity, bank size and financial stability proxied by NPL. The results indicate that increased liquidity in the banking sector improves financial stability. Bank size had a statistically significant positive relationship with NPLs. In other words, as bank size increases, exposure to credit risks increases. Interestingly, bank concentration was found to be inversely related to NPLs. That is, a rise in the level of concentration in banking assets tends to reduce NPLs in the banking industry. This finding corroborates the concentration-stability argument of Keeley (1990) who maintains that, higher levels of bank concentration may discourage excessive risk-taking behaviour by management of such banks. The result also confirms the argument of Park and Peristiani (2007), who maintain that bank concentration is mostly associated with higher franchise values, as such, shareholders of banks may reject investments that are risky and threaten their future profits. Liquidity was also statistically significant in explaining financial stability. We found a negative link between liquidity and NPL. The results show that an increase in liquidity reduces credit risks and ensures stability of the financial sector.

To identify the moderating role of concentration of banking assets in moderating the link between competition and stability of the banking sector in Ghana, the study employed a panel threshold regression to identify the threshold for concentration in the banking industry. The results of the threshold regression are presented in Columns (3) and (4) of Table 6.2. A threshold value of 5.3% was obtained for bank concentration. This implies that the effect of competition on bank Z-score changes when concentration in the banking industry goes beyond 5.3%. On the other hand, the effect of competition on NPL only changes when concentration in the banking industry goes beyond 16.8%. When the bank Z-score is used as the dependent variable (see Column 3), threshold value of asset concentration was identified to be 5.3%. This implies, that the effect of competition on stability of the banking industry changes after banking sector concentration exceeds 5.3%. In the case of the present study, we identified that the impact of competition (market power) on stability was stronger (18.23 units) and highly significant before the threshold value. The effect of competition on stability becomes weaker (5.6 units) and insignificant after the threshold value. The results imply that effect of competition on stability on Ghana's banking industry largely depends on the level of concentration of banking assets. The larger the concentration, the weaker the ability of competition to lead to improved stability.

The threshold value for concentration, when NPL is used as the outcome variable was found to be 16.8%. The effect of competition (market power) before the threshold value was found to be negative and significant. Thus, an increase in the level of market power result in a reduction of NPLs by 0.22 units. Though the effect of competition (market power) on NPL was still negative and significant after the threshold value, its impact had reduced to 0.09 units. These results imply that concentration influences the link involving competition and bank stability. High levels of concentration tend to dilute the effect of competition on bank stability.

6.5 Conclusions and Policy Implications

Despite several policies implemented since the Financial Sector Adjustment Programme of 1986 to liberalise Ghana's financial sector and enhance stability, stability of Ghana's banking sector continues to be a challenge. The recent takeover and consolidation of seven local banks have raised concerns about the stability of Ghana's Banking industry. The study therefore analysed the effect of competition on stability of Ghana's banking industry using data from 12 commercial banks in Ghana. These 12 banks together account for over 70% of assets in the banking industry and such could be used to mimic the entire banking industry in Ghana.

The findings of our study show that greater levels of market power (low competition) enhance stability of Ghana's financial sector irrespective of the measure of stability (whether Z-score or NPLs). This result is consistent with the competition-fragility argument of Keeley (1990) and Marcus (1984). The study also found that the effect of concentration on stability largely depends on how stability is measured. Concentration was found to have adverse effect on stability when the bank Z-score was used as a measure of stability whereas it was found to contribute to stability when NPL is employed as a stability measure. The differences in effect could be attributed to how concentration affects efficiency and risk-taking behaviour of banks. The study also found that higher levels of concentration irrespective of the measure of stability, tend to affect the extent to which competition (market power) affects stability. Specifically, concentration beyond 5.3% and 16.8% reduces stability effect of a rise in the level of market power for Bank Z-score and NPL, respectively.

The results of the study imply that the Bank of Ghana should work towards strengthening the capital adequacy framework and provide avenues for the consolidation of smaller sized banks using mergers and acquisitions. The consolidation of these banks would work to improve their level of market power, enhance their chances of survival in the market and the overall stability of the banking industry. In addition, some natural entry barriers (such as capital requirement to operate a bank) could be created to reduce competitive pressures in the industry since our empirical result provide evidence to the fact that lower level of competition is what enhances stability in the industry. It is important to also state that, in restricting entry into the banking industry, the Bank of Ghana must ensure that concentration of assets in the industry is well distributed.

These findings have several implications on stability of the banking sector for the post-COVID-19 pandemic periods. The Bank of Ghana playing its oversight responsibility should ensure that there is healthy competition among banks while monitoring the concentration of bank assets. Lack of proper monitoring of asset concentration may lead to dilution of the competition of banks. This in itself would render bank competition less effective in influencing the stability of the financial sector. Similarly, over concentration of bank assets may affect interest rates and credit making of the banks. In such instances, credit making by banks will be driven by profit motives of extremely larger banks. This may imply exclusion of access to needed finance for small enterprises and investment into a sector like agriculture. Such effect of bank asset concentration will hurt economic recovery of the country. Similarly, given that there is the likelihood of pressure on deposits and possibility of loans and financing boom due to COVID-19 pressures, there will be the need for banks to revamp their operations to meet the increasing demand. This will mean that banks would have to monitor deposits' variability, credit risk, loan and debt scheduling and refinancing with keen interest to ensure stability amidst growing competition. We also recommend that banks should consider a revision of any intended capital action (capital accumulation and investment) in line with prospects and challenges that present themselves in the current and post-pandemic era. Thus, banks must balance the act of capital investment, credit and liquidity in order to have asset quality and avoid liquidity constraints.

References

Adjei-Frimpong, K., Gan, C., & Hu, B. (2016). Competition in the banking industry: Empirical evidence from Ghana. *Journal of Banking Regulation, 17*(3), 159–175.

Agoraki, M. E. K., Delis, M. D., & Pasiouras, F. (2011). Regulations, competition and bank risk-taking in transition countries. *Journal of Financial Stability, 7*(1), 38–48.

Ahiabor, F. S. (2013). *Banking competition and profitability in Ghana: An empirical investigation of market power, bank-specific and macroeconomic determinants using system GMM* (MSc.). University of Leicester.

Alhassan, A. L., & Ohene-Asare, K. (2016). Competition and bank efficiency in emerging markets: Empirical evidence from Ghana. *African Journal of Economic and Management Studies, 7*(2), 268–288.

Amidu, M., & Wolfe, S. (2013). Does bank competition and diversification lead to greater stability? Evidence from emerging markets. *Review of Development Finance, 3*(3), 152–166.

Ariss, R. T. (2010). On the implications of market power in banking: Evidence from developing countries. *Journal of banking & Finance, 34*(4), 765–775.

Bank of Ghana. (2018). *Bank of Ghana annual report*. Accra.

Bank of Ghana. (2019). *Bank of Ghana annual report*. Accra.

Bank of Ghana. (2020). *Bank of Ghana annual report*. Accra.

Beck, T. (2008). *Bank competition and financial stability: Friends or foes?* The World Bank.

Berger, A. N., Klapper, L. F., & Turk-Ariss, R. (2017). Bank competition and financial stability. In *Handbook of competition in banking and finance*. Edward Elgar Publishing.

Boyd, J. H., & Runkle, D. E. (1993). Size and performance of banking firms: Testing the predictions of theory. *Journal of Monetary Economics, 31*(1), 47–67.

Buchs, T. D., & Mathisen, J. (2005). *Competition and efficiency in banking: Behavioral evidence from Ghana.*

Calice, P., & Leonida, L. (2018). *Concentration in the banking sector and financial stability: New evidence.* The World Bank.

Cetorelli, N., Hirtle, B., Morgan, D. P., Peristiani, S., & Santos, J. A. (2007). Trends in financial market concentration and their implications for market stability. *Economic Policy Review, 13*(1).

Clark, X., Dollar, D., & Micco, A. (2004). Port efficiency, maritime transport costs, and bilateral trade. *Journal of Development Economics, 75*(2), 417–450. https://doi.org/10.1016/j.jdeveco.2004.06.005

De Nicoló, M. G., Jalal, A. M., & Boyd, J. H. (2006). *Bank risk-taking and competition revisited: New theory and new evidence.* International Monetary Fund.

De-Ramon, S., Francis, W., & Straughan, M. (2018). *Bank competition and stability in the United Kingdom.*

Fernández, R. O., & Garza-García, J. G. (2015). The relationship between bank competition and financial stability: A case study of the Mexican banking industry. *Ensayos Revista de Economía (Ensayos Journal of Economics), 34*(1), 103–120.

Fu, X. M., Lin, Y. R., & Molyneux, P. (2014). Bank competition and financial stability in Asia Pacific. *Journal of Banking & Finance, 38*, 64–77.

Gudmundsson, R., Ngoka-Kisinguh, K., & Odongo, M. T. (2013). *The role of capital requirements on bank competition and stability: The case of the Kenyan banking industry.* Kenya Bankers Association-KBA Centre for Research on Financial Markets and Policy Working Paper Series.

Hainz, C., Weill, L., & Godlewski, C. J. (2013). Bank competition and collateral: Theory and evidence. *Journal of Financial Services Research, 44*(2), 131–148.

Hellmann, T. F., Murdock, K. C., & Stiglitz, J. E. (2000). Liberalization, moral hazard in banking, and prudential regulation: Are capital requirements enough? *American Economic Review, 90*(1), 147–165.

Horvath, R., Seidler, J., & Weill, L. (2016). How bank competition influences liquidity creation. *Economic Modelling, 52*, 155–161.

Kasman, S., & Kasman, A. (2015). Bank competition, concentration and financial stability in the Turkish banking industry. *Economic Systems, 39*(3), 502–517.

Keeley, M. C. (1990). Deposit insurance, risk, and market power in banking. *The American Economic Review, 80*, 1183–1200.

Koetter, M., Kolari, J. W., & Spierdijk, L. (2012). Enjoying the quiet life under deregulation? Evidence from adjusted Lerner indices for US banks. *Review of Economics and Statistics, 94*(2), 462–480.

Liu, H., Molyneux, P., & Wilson, J. O. (2013). Competition and stability in European banking: A regional analysis. *The Manchester School, 81*(2), 176–201.

Marcus, A. J. (1984). Deregulation and bank financial policy. *Journal of Banking & Finance, 8*(4), 557–565.

Martinez-Miera, D., & Repullo, R. (2010). Does competition reduce the risk of bank failure? *The Review of Financial Studies, 23*(10), 3638–3664.

Oduor, J., Ngoka, K., & Odongo, M. (2017). Capital requirement, bank competition and stability in Africa. *Review of Development Finance, 7*(1), 45–51.

Pak, O., & Nurmakhanova, M. (2013). The effect of market power on bank credit risk-taking and bank stability in Kazakhstan. *Transition Studies Review, 20*(3), 335–350.

Park, S., & Peristiani, S. (2007). Are bank shareholders enemies of regulators or a potential source of market discipline? *Journal of Banking & Finance, 31*(8), 2493–2515.

Petersen, M. A., & Rajan, R. G. (1995). The effect of credit market competition on lending relationships. *The Quarterly Journal of Economics, 110*(2), 407–443.

Pino, G., & Araya, I. (2013). Impact of the heterogeneity in market power on the relationship between risk taking and competition: Case of the Chilean banking sector. *Emerging Markets Finance and Trade, 49*(4), 98–112.

Ruza, C., de la Cuesta-González, M., & Paredes-Gazquez, J. (2019). Banking system resilience: An empirical appraisal. *Journal of Economic Studies, 46*(6), 1241–1257.

Saurina Salas, J., Jimenez, G., & Lopez, J. A. (2007, September). How does competition impact bank risk taking? In EFA 2007 Ljubljana Meetings Paper, FRB of San Francisco Working Paper (No. 2007–23).

Schaeck, K., Cihak, M., & Wolfe, S. (2009). Are competitive banking systems more stable? *Journal of Money, Credit and Banking, 41*(4), 711–734.

Uhde, A., & Heimeshoff, U. (2009). Consolidation in banking and financial stability in Europe: Empirical evidence. *Journal of Banking & Finance, 33*(7), 1299–1311.

Wang, Q. (2015). Fixed-effect panel threshold model using Stata. *The Stata Journal, 15*(1), 121–134.

Yaldiz, E., & Bazzana, F. (2010). The effect of market power on bank risk taking in Turkey. *Financial Theory and Practice, 34*(3), 297–314.

Yeyati, E. L., & Micco, A. (2007). Concentration and foreign penetration in Latin American banking sectors: Impact on competition and risk. *Journal of Banking & Finance, 31*(6), 1633–1647.

CHAPTER 7

Macroeconomic Determinants of Banking Instability in Ghana

Harold Ngalawa and Evelyn Derera

7.1 Introduction

Modern economies cannot function effectively without an efficient banking system intermediating between surplus and deficit spending units and providing other essential financial services (see Ngari & Muiruri, 2014; Mertens & Urga, 2001). In Ghana, the banking sector constitutes 23 banks (2021 estimate) (see Appendix 1 for a complete list of registered commercial banks in Ghana), 144 rural and community banks, 137 deposit-taking microfinance institutions, 25 savings and loans companies and 31 microcredit institutions, among others (see Bank of Ghana,

H. Ngalawa (✉)
School of Accounting, Economics and Finance, University of KwaZulu-Natal, Durban, South Africa
e-mail: Ngalawa@ukzn.ac.za

E. Derera
School of Management, IT & Governance, University of KwaZulu-Natal, Durban, South Africa
e-mail: dererae@ukzn.ac.za

J. A. Peprah et al. (eds.), *Financial Sector Development in Ghana*, Palgrave Macmillan Studies in Banking and Financial Institutions,
https://doi.org/10.1007/978-3-031-09345-6_7

2022). According to PwC (2021), a significant share of the commercial banking market in Ghana is controlled by the country's six largest banks consisting of Ecobank Ghana Limited, GCB Bank Limited, Stanbic Bank Ghana Limited, ABSA Bank Ghana Limited, Consolidated Bank Ghana Limited and Fidelity Bank Ghana Limited. These banks account for 51.9% of total assets (2021 estimate) (see PwC, 2021). On the other hand, the six smallest commercial banks in the country account for only 8.9% of the total assets in the commercial banking sector (PwC, 2021). These are First Atlantic Bank Limited, National Investment Bank Limited, First National Bank (Ghana) Limited, Bank of Africa Ghana Limited, FBN Bank (Ghana) Limited and OmniBSIC Bank Ghana Limited (see PwC, 2021).

During the period 2007–2011, total banking assets in Ghana grew at a rate of 28.2% per annum (see Aithnard et al., 2013). According to Aithnard et al. (2013), total assets in Ghana's banking sector amounted to US$13.39 billion by 2011, partly as a result of regulations that led to capital raising programmes, and partly due to an economic boom propelled by the development of the oil and gas sectors. While total assets increased, the ratio of loans to total assets decreased from 52.4% to 34.5% during the same period, suggesting that the increase in total assets was not necessarily driven by rising credit; and by 2016, the growth of credit to the private sector had declined to 9% from 36.65% in 2015 (see Aithnard et al., 2013).

During the period 1976–1983, Ghana experienced deteriorating economic growth and severe balance of payments deficits (Kamasa et al., 2020). In 1987, the Government of Ghana, in liaison with the World Bank and International Monetary Fund (IMF), embarked on financial sector reforms within the framework of a Financial Sector Adjustment Programme (see Obuobi et al., 2019), which transformed the financial sector from a repressed to a highly liberalised segment of the economy (Kamasa et al., 2020). Other reforms included the Economic Recovery Program in 1983 and revisions of the Bank of Ghana Act in 2002, Banking Amendment Act in 2007, Foreign Exchange Act in 2007, Credit Reporting Act in 2008 and Lender and Borrowers Act in 2008, inter alia (see Kamasa et al., 2020).

Ghana has had several conspicuous incidents of banking instability. In August 2017, the Bank of Ghana announced the takeover of UT Bank Limited and Capital Bank Limited by CBG Limited, citing insolvency as a reason for the failure of the two banks (Bank of Ghana, 2017). A year later, in August 2018, the Bank of Ghana announced that it had revoked licences of five banks, namely, uniBank Limited, the Royal Bank Limited, Beige Bank Limited, Sovereign Bank Limited and Construction Bank Limited, also citing insolvency as a reason for their closure (Bank of Ghana, 2018). All assets and liabilities of the five banks were transferred to Consolidated Bank Ghana (CBG) Limited, a new bank established specifically to take over operations of the five banks.

Several scholars have investigated factors that may explain banking instability in Ghana. One of the distinguishing features among these studies is the measure of banking instability employed. Among others, Moyo et al. (2014), Kong et al. (2020), Antwi (2020) and Yensu et al. (2021) use accounting-based bank-level Z-scores to measure banking stability. The Z-score is inversely related to the probability of bank insolvency. A larger value of this measure indicates less overall bank risk and higher bank stability. While the Z-scores are widely used, they provide only a single measure of banking instability: the insolvency problem of banks. This study contributes to the literature by using a relatively new empirical framework that distinguishes between banking instability caused by bank runs or panic withdrawals of deposits, and banking instability emanating from the insolvency problem of banks. Unlike the previous studies (see for example, Moyo et al., 2014; Kong et al., 2020; Antwi, 2020), we do not assume that banking instability can only be characterised by bank insolvency. To the extent that both bank runs and bank insolvency are used separately to construct distinct measures of bank instability, this study achieves the depth that has not been attained by the aforementioned and similar studies on Ghana.

Findings of this study are expected to guide policymakers in Ghana on appropriate policy formulation and implementation aimed at achieving banking sector stability. In a well-functioning financial sector, banks mobilise financial resources from surplus spenders and direct them towards the pursuit of efficient economic activities carried out by deficit spending units. Consequently, this reduces the cost of transferring funds from borrowers to lenders, which leads to a more efficient allocation of resources and faster economic growth. In the literature, there is near consensus that well-functioning banks accelerate economic growth, while poorly functioning banks are an obstacle to economic progress and aggravate poverty (see Drigă & Dura, 2014). If policymakers are able to identify factors that lead to banking instability, they can contribute to economic growth and poverty reduction by formulating and implementing well-informed policies that lead to an efficient and stable banking sector.

This chapter sets out to investigate the macroeconomic factors that explain banking instability in Ghana. The study constructs indices of bank runs and insolvency of banks, separately, as distinct proxies of banking instability; and employs quarterly frequency time series data for the period 1996 to 2020 and an ordered logistic model to carry out the analysis. We categorise the macroeconomic fundamentals into three groups, namely, macroeconomic factors that a priori are expected to affect the quality of bank assets (growth of real gross domestic product, inflation and real interest rates); factors that are expected to capture the exposure of local banks to foreign exchange risk (exchange rate depreciation/appreciation and the ratio of aggregate money supply (M2) to foreign exchange reserves); and factors that capture the quality of regulatory and institutional structures in the country (political stability and the rule of law).

Following this introduction, the rest of the chapter is organised in four sections. Section 2 presents a theoretical discussion of banking instability. Methodology, data and data sources are outlined in Sect. 3, while estimation results and inferences follow in Sect. 4. Section 5 presents a summary, conclusions and policy implications.

7.2 Banking Instability: A Review of the Literature

Banking instability can be characterised by a bank run (panic withdrawal of deposits) or bank insolvency (Ngalawa, 2012). A bank run or panic withdrawal of deposits occurs when depositors respond to news and hence expectations of bank failure by rushing to withdraw their deposits in full (Ngalawa et al., 2016). If this action by the depositors persists, the bank may be forced to liquidate its assets at a loss, consequently leading to a collapse (failure) of the bank (see Ngalawa, 2012; Ngalawa et al., 2016).

There are several explanations of possible trigger mechanisms of banking instability characterised by bank runs. One of the earliest accounts, presented by Fisher (1911), maintains that a bank run occurs whenever a bank's total deposits (liabilities) exceed the value of its assets, which forces depositors to rush to withdraw their deposits in full in order to minimise their losses (see also Grevenitis et al., 2020). Chari and Jagannathan (1988) have argued that a run on a bank can occur even if there is no one with adverse information about the operations or future returns of the bank. In Chari and Jagannathan's (1988) model, individuals observe long queues at a bank but not the information content held by the people on the queues. The individuals observing the long queues conclude that the bank is on the verge of failing, regardless of the information content held by the people on the queues, which consequently precipitates a run on the bank. Thus, a bank run in the Chari and Jagannathan (1988) framework is modelled as an equilibrium position in a formulation where all equilibria have bank runs.

An alternative explanation is provided in Diamond and Dybvig's (1983) canonical study, which holds that a bank run is caused by a shift in expectations, which may depend on almost anything, referred to as sunspots run equilibrium (Adao & Temzelides, 1998; Carmona, 2004; Diamond & Dybvig, 1983). In this model, a bank run is presented as a bad equilibrium in a series of possible multiple equilibria.

Banking instability can also be characterised by the insolvency of banks, usually represented by large amounts of unexpected non-performing loans. Caprio and Klingebiel (1997) define insolvency as a case where the net worth of the banking system has been entirely (or nearly) eliminated. This situation is highly likely to occur during an economic downturn and after a period of a boom in private sector lending (Allen & Gale, 1998; Caprio & Klingebiel, 1997).

In a period of banking instability characterised by the insolvency problem of banks arising from large amounts of non-performing loans, banks tend to change their lending policy in a bid to stop this negative pattern that has the potential to lead to bank failure. While redrafting their standard operating procedures and restructuring their operations, the banks are somewhat cautious in their business approach, which may explain the significant drop in their credit to the private sector. In the event that a bank has been taken over by new management or by government authorities following an insolvency problem, a period of low credit extension to the private sector follows during the restructuring phase, particularly because of illiquidity problems and the need to adopt a safe credit policy.

There are two predominant theoretical views on factors that explain banking instability, namely, the self-fulfilling view and the fundamental view (see Foley-Fisher et al., 2020). The self-fulfilling view holds that banking instability is a realisation of a bad equilibrium arising from self-fulfilling expectations in a multiple equilibria framework (see Foley-Fisher et al., 2020). The fundamental banking instability view, on the other hand, perceives banking instability as a consequence of poor economic performance (see Allen & Gale, 1998).

7.3 Methodology, Data and Data Sources

7.3.1 *Data and Data Sources*

The study employs quarterly frequency time series data for Ghana covering the period 1996–2020. The cut-off dates were dictated by data availability. Data sources include the World Development Indicators (WDI) (a World Bank database of socioeconomic and demographic indicators), Worldwide Governance Indicators (WGI) (a World Bank database on the quality of governance in industrial and developing countries), International Financial Statistics (IFS) (an International Monetary Fund (IMF) database), the Bank of Ghana (BOG) and Ghana Statistical Services (GSS) (the National Statistical Office of Ghana).

7.3.2 *Measures of Banking Instability*

We measure banking instability using two alternative indices, namely index of bank runs and the index of bank insolvency. Following von Hagen and Ho (2007) and Eichengreen, Rose and Wyplosz (1995, 1996a and 1996b), we use monthly frequency data to compute an index of bank runs (IBR_t) using deseasonalised growth rates of demand deposits (GDD_t) and time deposits (GTD_t), and an index of bank insolvency (ISV_t) using deseasonalised growth rates of credit extended by commercial banks to the private sector (GPC_t) (see also Ngalawa, 2012; Ngalawa et al., 2016). We commence by calculating the deseasonalised growth rates of each of the three series as follows:

$$GDD_t = \frac{(DD_t - DD_{t-12})}{DD_{t-12}} \tag{7.1}$$

$$GTD_t = \frac{(TD_t - TD_{t-12})}{TD_{t-12}} \tag{7.2}$$

$$GPC_t = \frac{(PC_t - PC_{t-12})}{PC_{t-12}} \tag{7.3}$$

where DD, TD and PC represent demand deposits, time deposits and commercial bank credit to the private sector, in that order, and t is a time index. Deseasonalisation (also referred to as seasonal adjustment) of data is a statistical technique for removing cyclical seasonal movements that are common in time series, while retaining the underlying trend component. It is imperative to deseasonalise the data in order to understand the underlying trends and cycles. Next, we compute indices of bank runs and bank insolvency as follows:

$$IBR_t = \left[\left(\frac{GDD_t - \overline{GDD_t}}{\sigma_{GDD_t}}\right) + \left(\frac{GTD_t - \overline{GTD_t}}{\sigma_{GTD_t}}\right)\right] \Big/ 2 \tag{7.4}$$

$$ISV_t = \left(\frac{GPC_t - \overline{GPC_t}}{\sigma_{GPC_t}}\right) \tag{7.5}$$

where $\overline{GDD_t}$, $\overline{GTD_t}$ and $\overline{GPC_t}$ are mean growth rates of demand deposits, time deposits and commercial bank credit to the private sector, respectively; and σ_{GDD_t}, σ_{GTD_t} and σ_{GPC_t} are standard deviations of the deseasonalised growth rates of demand deposits, time deposits and commercial bank credit to the private sector, in that order.

Since bank runs and bank insolvency are both characterised by sharp decreases in bank deposits and domestic private sector credit, respectively, we use extreme values of *IBR* and *ISV* as measures of bank runs and insolvency of banks. We distinguish between broad and narrow definitions of banking instability. With the broad measure of banking instability, we characterise the indices that fall within the lowest 10% of the standard normal distribution as a representation of banking instability, and they take the value 1. If the calculated indices fall within the highest 90% of the standard normal distribution, we classify them as indicators of banking stability, and the measure takes the value zero. Analogously, the narrow measure of banking instability denotes the calculated indices that fall within the lowest 5% of the standard normal distribution as indicators of banking instability and they take the value 1. If the indices fall within the highest 95% of the standard normal distribution, they are classified as indicators of banking stability, and they take the value zero. We denote these measures of bank runs and insolvency of banks as RUN_t and SLV_t, in that order.

While the indices are computed at monthly frequency, analysis is carried out at quarterly frequency. The banking instability variable gets the value zero if there is no observation of banking instability in a quarter (period of banking stability); 1 if there is only one observation of banking instability in a quarter; 2 if there are two observations of banking instability in a quarter; and 3 if there are three observations of banking instability in a quarter (period of severe banking instability).

7.3.3 Measures of Macroeconomic Performance

We use seven variables to measure macroeconomic performance, namely, growth of real gross domestic product (GDP)($ggdp$), the all items national consumer price inflation($infl$), interest rates(ir), exchange rate depreciation/appreciation($cexr$), the ratio of aggregate money supply (M2) to foreign exchange reserves($msfx$), political stability (pol) and the rule of law(rol). These indicators are classified into three categories. The first three variables (consumer price inflation, real GDP growth and interest rates) are macroeconomic factors that a priori are expected to affect the quality of bank assets. The next two variables (exchange rate depreciation/appreciation and the ratio of aggregate money supply (M2) to foreign exchange reserves) are expected to capture the exposure of local banks to foreign exchange risk. The last two variables (political stability and the rule of law) are postulated to capture the quality of regulatory and institutional structures in the country.

Real GDP growth is a reflection of an increase in the level of income. This enhances the ability of borrowers to service their loans, which in turn improves the quality of bank assets and consequently minimises the probability of bank runs. Real GDP growth, therefore, is expected to be inversely related to banking instability. GDP growth is measured as year-on-year percentage changes in real GDP (base year 2010). The data are obtained from the World Bank's World Development Indicators.

It is expected that the relationship between inflation and bank instability is non-linear. High rates of inflation reflect escalating operating costs, which in turn reduce the ability of borrowers to service their loans, leading to a rise in non-performing loans and, subsequently, an increase in the probability of insolvency of banks. Some studies have argued that the high inflation rates cause uncertainty that compels economic agents to protect themselves from the rising price instability, leading to a slowdown in economic activity, which tends to impair the agents' ability to repay their loans, subsequently raising the probability of banking instability (see for example, Li, 2006; Ghosh & Phillips, 1998). News that banks are accumulating large amounts of non-performing loans create expectations among depositors that the bank is about to fail. In this event, depositors respond by rushing to withdraw their deposits in full, forcing the bank to liquidate its assets at a loss, and leading to bank failure (self-fulfilling expectations). It is, therefore, expected that high inflation and banking instability are positively related.

If prices are relatively stable (low inflation), on the other hand, economic agents operate in a near-certainty environment for their long-term expenditure plans, and have access to cheaper credit, which increases investment, consumption and output (see Brook et al., 2002; Almeida et al., 2004; Morar, 2011), and hence the ability to repay their loans. Consequently, the probability of banking stability increases as well. Thus, low inflation and banking instability are inversely related. For purposes of this study, inflation is measured by year-on-year percentage changes in the all items national composite consumer price index (CPI). The data were collected from the Bank of Ghana website (https://www.bog.gov.gh).

An increase in interest rates entails a rise in the cost of borrowing. Borrowers with variable interest rate contracts are expected to be adversely affected by an interest rate hike. The higher interest rates are associated with a loss in income equivalent to the difference between what is actually paid to service the loan and what would have been paid if the interest rates had remained low. This is expected to negatively affect the borrowers' cash flow, consequently constraining the ability to service their loans, and leading to high default rates. On the part of the lending banks, high interest rates appear on the bank's balance sheet as an increase in the cost of funds (e.g. deposits). Since banks usually fix interest rates on bank loans and other assets over relatively long periods of time, the higher interest rates increase the cost of funds, which ultimately has a negative effect on the liability side of the bank's balance sheet. Therefore, whether viewed from the lenders' (banks') or borrowers' point of view, interest rates and the probability of banking instability are expected to change together in the same direction. Interest rates are measured as period average savings deposit rates. The data were collected from the Bank of Ghana website (https://www.bog.gov.gh).

A depreciation or an appreciation of the exchange rate and the ratio of aggregate money supply (*M2*) to foreign exchange reserves are used to capture commercial banks' vulnerability to sudden capital outflows triggered by a run on the currency and the banks' exposure to foreign exchange risk (see Demirgüç-Kunt & Detragiache, 2002). A depreciation of the domestic currency against major trading currencies is expected to trigger withdrawals of domestic currency deposits to purchase foreign assets, and vice versa. The larger the extent to which this takes place, the more vulnerable are the banks to foreign exchange risk. If the exchange rate is measured using the direct approach, the higher the extent of the currency's depreciation, the higher the probability of banking instability. We measure the exchange rate in Ghanaian Cedis per US dollar, and the data were collected from the World Bank's World Development Indicators. Data on aggregate money supply (M2) as a ratio of foreign exchange reserves are also obtained from the World Development Indicators compiled by the World Bank.

It is important to note that the impact of exchange rate risk on banking sector stability is dependent on the level of capital account openness and the amount of foreign capital in the banking industry. Ishii et al. (2002) argue that if foreign currency exposure (of banks) is large enough, it may lead to insolvency or liquidity problems that may trigger a chain reaction that can give rise to systemic banking instability.

The study also employs two variables, namely "political stability (and absence of terrorism)" and "rule of law" to capture the institutional as well as the regulatory environment in Ghana. Political stability and the absence of violence/terrorism measures "perceptions of the likelihood of political instability and/or politically-motivated violence, including terrorism". The "rule of law", on the other hand, "reflects perceptions of the extent to which agents have confidence in and abide by the rules of society, and in particular the quality of contract enforcement, property rights, the police and the courts, as well as the likelihood of crime and violence" (see Kaufmann et al., 2010). Data for both variables were collected from the Worldwide Governance Indicators compiled by the World Bank. The data were compiled from surveys as well as views from public, private and non-governmental organisation experts on governance. An estimate of governance ranges from approximately −2.5 (weak governance) to 2.5 (strong governance). A priori, it is expected that a weak institutional and regulatory environment is associated with banking instability while a strong institutional and regulatory environment is associated with banking instability.

Besides inflation rates, exchange rates and interest rates, the rest of the variables are only available in annual frequency. Therefore, we interpolated these variables from annual to quarterly frequency using the quadratic average match frequency conversion technique. In this method, "a quadratic polynomial is formed by taking three adjacent points from the source series and fitting a quadratic so that either the average or the sum of the high frequency points matches the low frequency data actually observed" (Global, 2017).

7.3.4 The Logit Model

Financial instability takes the values 0, 1, 2 and 3, where 0 represents financial stability; 1 and 2 characterise moderate financial instability; and 3 denotes severe banking instability. The study, therefore, adopts an ordinal regression model for analysis. There are two major specifications of ordinal regression models, namely, the logit and probit. According to Gujarati (2003) and Jacobs and Kuper (2007), the predicted probabilities computed by these two models differ only slightly. In addition, the technical differences as well as output disparities from the two models are minor.

This study estimates an ordered logistic model (also referred to as a proportional odds model). This is a regression model for an ordinal response variable, which is based on the cumulative probabilities of the response variable (Grilli & Rampichini, 2014). To illustrate the mechanics of the model, we commence by postulating that the probability of financial instability in Ghana [$\Pr(Inst_t \epsilon (RUN_t, SLV_t))$] is given by:

$$\Pr(\text{Inst}_t) = \text{E}_t(\text{Inst}_t = 1|\text{X}_t) = \frac{e^{(\alpha_t+\beta_t X_t+\nu_t)}}{1 + e^{(\alpha_t+\beta_t X_t+\nu_t)}} \tag{7.6}$$

$$\text{where Inst}_t = \begin{cases} 3 \text{ if there is severe finacial instability in Ghana at time t} \\ 1 \text{ and } 2 \text{ if there is mild financial instability in Ghana at time t} \\ 0 \text{ if there is financial stability in Ghana at time t} \end{cases}$$

X_t is a vector of macroeconomic indicators, α and β are parameters, and ν_t is a random error term. It can be shown that the logit specification in Eq. (7.6) reduces to:

$$\frac{\Pr(\text{Inst}_t)}{1 - \Pr(\text{Inst}_t)} = e^{(\alpha_t+\beta_t X_t+\nu_t)} \tag{7.7}$$

which can be rewritten as:

$$L_t = \ln \frac{\Pr(\text{Inst}_t)}{1 - \Pr(\text{Inst}_t)} = \alpha_t + \beta_t X_t + \nu_t \tag{7.8}$$

In Eq. (7.8), L_t is the regressand or logit. The empirical regression equation, therefore, is given by:

$$\text{Inst}_t = \beta_o + \sum_{i=1}^{n} \beta_i Z_{it} + \mu_t \tag{7.9}$$

where $Z_{it} = X_t \epsilon (\text{ggdp}_t, \text{infl}_t, \text{infl}_t^2, \text{ir}_t, \text{cexr}_t, \text{msfx}_t, \text{pol}_t, \text{rol}_t)$
Equation (7.9) can also be rewritten as:

$$\begin{aligned}\text{Inst}_t = {} & \beta_o + \beta_1 \text{ggdp}_t + \beta_2 \text{infl}_t + \beta_3 \text{infl}_t^2 + \beta_4 \text{ir}_t \\ & + \beta_5 \text{cexr}_t + \beta_6 \text{msfx}_t + \beta_7 \text{pol}_t + \beta_8 \text{rol}_t + \mu_t\end{aligned} \tag{7.10}$$

7.4 Study Results and Inferences

7.4.1 Descriptive Statistics

Table 7.1 presents descriptive statistics for the macroeconomic variables used in the study. The table shows that during the study period (1996Q1–2020Q4), GDP in Ghana grew by an average of 5.6% with the lowest recorded at −3.2% (in the fourth quarter of 2020) and the highest recorded at14.47% (in the third quarter of 2011) (see also Fig. 7.1). The lowest GDP growth during the study period occurred in the third quarter of 2020, at a time when nearly the whole world (Ghana included) was adversely affected by the COVID-19 pandemic. The rate of inflation in the country was relatively high and volatile during the study period as reflected by the comparatively high mean of 17.4% and standard deviation of 6.8%. Inflation in Ghana varied widely between 1996 and 2020, with the minimum inflation rate recorded at 7.8% (in the first quarter of 2020) and the highest recorded at 67.3% (in the first quarter of 1996). Overall, the rate of inflation has been on a downward trend during the study period (see Fig. 7.1).

Table 7.1 further shows that interest rates in Ghana varied from a low of 4.1% (in the first quarter of 2012) to a high of 27.7% (in the first quarter of 1996) with a standard deviation of 6.8% during the sample period. Interest rates in the country were highest between 1996 and 1997 when they averaged 27.5% (see also Fig. 7.1). On the whole, the country has experienced significantly declining interest rates during the study period.

Table 7.1 Descriptive statistics

Variable	*Observations*	*Mean*	*Standard Deviations*	*Minimum*	*Maximum*
GDP growth	100	5.5966	2.7897	−3.2330	14.4749
CPI Inflation	100	17.3882	10.0058	7.8000	67.3400
Interest rates	100	10.4953	6.8396	4.0500	27.6500
Change in exchange rates	100	1.8769	1.6634	0.1448	5.6821
M2 as a ratio of foreign exchange reserves	100	2.6094	0.8552	1.1047	4.6887
Political stability	100	−0.0388	0.1458	−0.3673	0.1812
Rule of law	100	−0.0056	0.1206	−0.3047	0.1760

The exchange rate of Ghanaian Cedis (GHS) per US dollar (US$) varied from a minimum of GHS0.1448 per US$ (in the first quarter of 1996) to a maximum of GHS5.6821 per US$ (in the fourth quarter of 2020). The exchange rate averaged GHS1.8769 per US$ with a standard deviation of GHS0.1448 per US$ during the study period. Finally, Table 7.1 reveals that the ratio of aggregate money supply to foreign exchange reserves averaged 2.6, varying between a low of 1.1 and 4.7.

7.4.2 Unit Root Test

We carry out a test for stationarity using the Phillips-Perron Test to establish the order of integration for each of the variables (see Table 7.2 for the unit root test results). If the P-value for the Phillips-Perron Test statistic of a variable is less than 0.01, 0.05 or 0.1, then the variable is stationary (i.e. has no unit roots) at 1%, 5% or 10% critical values, in that order. Table 7.2 shows that inflation and change in exchange rates (depreciation or appreciation of the Ghanaian Cedi vis-à-vis the US dollar) are stationary in levels (i.e. integrated of order zero), while GDP growth, interest rates and the ratio of aggregate money supply to foreign exchange reserves are stationary in first differences (i.e. integrated of order 1).

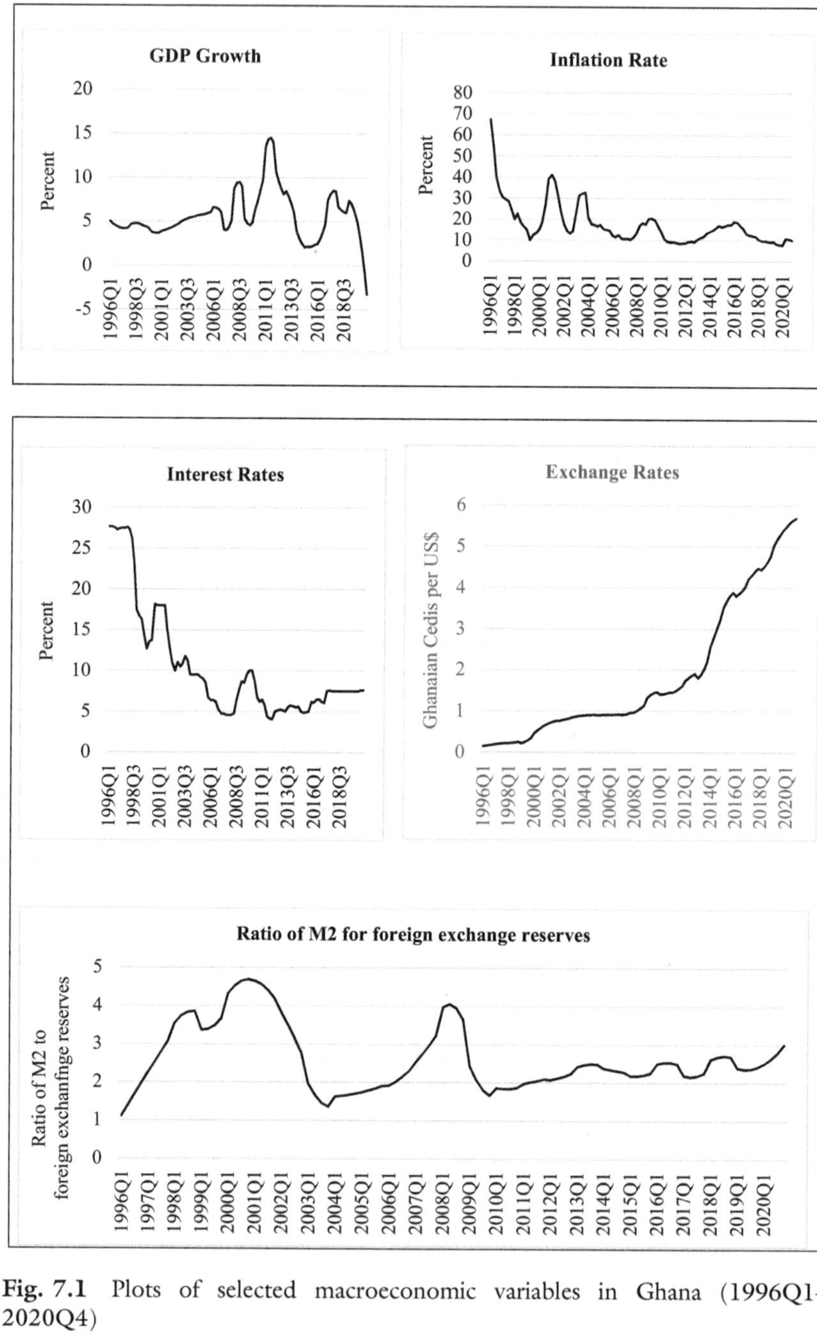

Fig. 7.1 Plots of selected macroeconomic variables in Ghana (1996Q1–2020Q4)

Table 7.2 Phillips-Perron Results for Stationarity

Variable	*Levels*		*First differences*		*Order of integration*
	PP t-statistic	*P-Value*	*PP t-statistic*	*P-Value*	
GDP growth	−1.6966	0.7456	−5.0001	0.0005	I(1)
Inflation	−5.4952	0.0001			I(0)
Interest rates	−1.8152	0.6901	−6.6220	0.0000	I(1)
Change in Exchange rates	−4.1447	0.0013			I(0)
M2 as a ratio of foreign exchange reserves	−2.9385	0.1552	−4.8488	0.0008	I(1)

NB Test critical values: *,**,*** represent 10%, 5% and 1%, respectively
Source Computed by the authors

7.4.3 Estimation Results

Table 7.3 presents estimation results of banking instability measured by bank runs broadly defined. Among the macroeconomic factors that are expected to affect the quality of bank assets, the table shows that the probability of banking instability decreases with an increase in real output growth; increases with a rise in interest rates; and has a non-linear relationship with inflation rates. The finding that changes in GDP growth are inversely related to banking instability agrees with a priori theoretical expectations. This confirms that as the level of income increases, borrowers' ability to service their loans improves, which in turn enhances the quality of the banks' asset base, and consequently reduces the probability of bank runs.

The results in Table 7.3 also show that inflation rates have a curvilinear relationship with banking instability characterised by bank runs broadly defined. The table reveals that an increase in inflation reduces the probability of banking instability while an increase in inflation squared increases the probability of banking instability. This shows that at low to moderate levels of inflation, there is an inverse relationship between banking instability and inflation rates, while at high levels of inflation, there is a positive relationship between inflation and banking instability. According to Morar (2011), Almeida (2004) and Brook et al. (2002), at low inflation rates, interest rates are also relatively low. Accordingly, economic agents tend to have access to cheap credit, which increases real economic activity, consequently increasing the agent's ability to repay their loans. Other studies have also argued that at low levels of inflation, economic agents face a near-certainty environment for their long-term expenditure plans, which increase real investment and output (see Morar, 2011, Almeida, 2004 & Brook et al., 2002), subsequently increasing the agents' ability to repay their loans, and decreasing the probability of bank stability.

High inflation rates, on the other hand, reflect rising operating costs, which in turn reduce the ability of borrowers to service their loans, leading to a rise in non-performing loans (see International Monetary Fund, 2011; Adusei, 2015). As news that a bank is accumulating large amounts of non-performing loans spreads, the chance of panic withdraws of deposits (bank runs) increases. As a result, the probability of bank instability also increases. According to Li (2006) and Ghosh and Phillips (1998), high inflation rates are also associated with rising uncertainty that forces economic agents to protect themselves against the rising general price level. This leads to a decline in real economic activity and hence income, which impairs the economic agents' ability to repay their loans. Consequently, the probability of banking instability increases.

Table 7.3 further shows that a rise in interest rates increases the probability of banking instability characterised by bank runs broadly defined.

Table 7.3 Regression results of banking instability measured by bank runs broadly defined

Dependent variable: Bank run broadly defined			
Variable	Coefficient	Standard Error	P-Value
GDP growth	−0.4786**	0.2261	0.0340
CPI inflation	−0.2417***	0.0900	0.0070
CPI inflation squared	0.0030*	0.0017	0.0860
Interest rates	0.7923***	0.2920	0.0070
Change in exchange rates	16.1552***	6.2243	0.0090
M2 as a ratio of foreign exchange reserves	2.0392*	1.0685	0.0560
Political stability	0.4657	2.9915	0.8760
Rule of law	−4.0966*	2.4656	0.0970
Number of observations = 92			
Wald chi2 (8) = 33.84			
Prob > chi2 = 0.0000			

NB Test critical values: *,**,*** represent 10%, 5% and 1%, respectively
Source Computed by the authors

Borrowers tend to be adversely affected by high costs of borrowing when interest rates go up. High interest rates reflect a loss in income, with negative implications on the borrower's cash flow leading to high default rates. On the part of banks, high interest rates increase the cost of funds, which negatively affect the liability side of their balance sheets. Either way, the probability of bank runs goes up as interest rates increase.

Table 7.3 also shows that exchange rates and the probability of banking instability are positively related. This indicates that a depreciation of the Ghanaian Cedi vis-à-vis the US dollar increases the probability of banking instability measured by bank runs broadly defined. This may be explained by withdrawals of domestic currency deposits to purchase foreign assets, which are seen as a relatively safer haven in comparison to the local currency, following expectations of a depreciation in the Ghanaian Cedi against major trading currencies (including the US dollar). The greater the extent to which this takes place, the more vulnerable are the banks to foreign exchange risk. In the end, we observe a positive relationship between banking instability and exchange rate depreciation.

Among the two variables that are expected to capture the exposure of local banks to foreign exchange risk, we find evidence that a depreciation of the Ghanaian Cedi vis-à-vis the US dollar and a rise in the ratio of aggregate money supply (M2) to foreign exchange reserves increase the probability of banking instability characterised by bank runs broadly defined (see Table 7.3). This indicates that purchases and sales of foreign assets may trigger banking instability in Ghana. A depreciation of the Ghanaian Cedi against the US dollar is expected to trigger withdrawals of domestic currency deposits to purchase foreign assets, and the extent to which this takes place determines the magnitude of banking instability in the country.

The estimation results in Table 7.3 further show that there is a positive relationship between the ratio of aggregate money supply (M2) to foreign exchange reserves and banking instability in Ghana measured by bank runs broadly defined. This suggests that an increase in commercial bank exposure to foreign exchange risk is associated with banking instability in Ghana.

Two variables are used in the model to capture the quality of regulatory and institutional structures in Ghana. These are political stability (and absence of terrorism) and the rule of law. The study finds no evidence of a significant relationship between political stability (as well as the absence of terrorism) and banking instability in the form of bank runs broadly defined (see Table 7.3). It is, however, observed that the rule of law and banking instability characterised by bank runs broadly defined are inversely related. This means that the banking system becomes more stable with improvements in perceptions of the extent to which economic agents have confidence in the laws governing the country, and specifically the quality of contract enforcement, property rights, the police and the courts (see Kaufmann et al., 2010).

Table 7.4 presents estimation results of banking instability characterised by bank runs narrowly defined. The primary difference between the estimation results in Table 7.3 and Table 7.4 is that the former uses observations (of bank run indices) in the lowest 10%, while the latter uses observations in the lowest 5% of the standard normal distribution as measures of banking instability.

The results in Table 7.4 corroborate the findings in Table 7.3. The table reveals that a rise in economic growth reduces the probability of banking instability characterised by bank runs narrowly defined. It is also observed that inflation has a non-linear relationship with the probability of bank runs narrowly defined. At low to moderate levels, a rise in inflation leads to a decline in the probability of bank runs narrowly defined, while at high rates, an increase in inflation also increases the probability of banking instability measured by bank runs narrowly defined.

Table 7.4 Regression results of banking instability measured by bank runs narrowly defined

Dependent variable: Bank run narrowly defined

Variable	Coefficient	Standard Error	P-Value
GDP growth	−0.5694***	0.2059	0.0060
CPI inflation	−0.1513***	0.0471	0.0010
CPI inflation squared	0.0019*	0.0011	0.0880
Interest rates	0.8381***	0.2931	0.0040
Change in exchange rates	21.6048**	9.9762	0.0300
M2 as a ratio of foreign exchange reserves	3.3874**	1.5973	0.0340
Political stability	−1.1116	2.7276	0.6840
Rule of law	−7.7624*	3.9744	0.0510
Number of observations = 86			
Wald chi2 (8) = 42.44			
Prob > chi2 = 0.0000			

NB Test critical values: *,**,*** represent 10%, 5% and 1%, respectively
Source Computed by the authors

It is further established that interest rates, exchange rates and the ratio of broad money to foreign exchange reserves change together with bank instability in the same direction. Thus, an increase in interest rates, a depreciation of the Ghanaian Cedi vis-à-vis the US dollar and a rise in the ratio of aggregate money supply to foreign exchange reserves decrease the probability of banking instability characterised by bank runs narrowly defined.

Table 7.4 also shows no evidence of a causal relationship between bank runs narrowly defined and political stability (and absence of terrorism). There is, however, an inverse relationship between the rule of law and bank runs narrowly defined. This reveals that the probability of banking instability in Ghana characterised by bank runs narrowly defined increases with a deterioration in the country's rule of law reflected by increasingly negative perceptions of the extent to which agents have faith in and follow societal rules, and especially the quality of contract enforcement, property rights, the police and the courts and the likelihood of crime and violence (see Kaufmann et al., 2010).

Table 7.5 presents estimation results of banking instability measured by the insolvency problem of banks broadly defined; and Table 7.6 presents regression results of banking instability depicted by the insolvency problem of banks narrowly defined. The results from the two tables are similar for all variables. Tables 7.5 and 7.6 show that the growth of real output decreases the probability of bank insolvency. This finding is similar to the relationship between changes in real GDP growth and banking instability measured by bank runs, both narrowly and broadly defined (see Tables 7.3 and 7.4). Thus, an increase in the growth of real income improves the borrowers' ability to service their loans, which in turn enhances the quality of bank assets and consequently minimises the probability of bank insolvency. Real GDP growth, therefore, is expected to be inversely related to banking instability.

Tables 7.5 and 7.6 also show that inflation has a curvilinear relationship with banking instability, which is consistent with the relationship between inflation rates and bank runs, both narrowly and broadly defined as reported in Tables 7.3 and 7.4. While an increase in low to moderate inflation reduces the probability of bank insolvency, a rise in high rates of inflation tends to increase the probability of bank insolvency, both narrowly and broadly defined (see Tables 7.5 and 7.6).

If inflation is relatively low, economic agents tend to have access to cheaper credit, leading to an increase in investment, output and hence income. Accordingly, the agents' ability to repay their loans increases. The effect on the banks is that non-performing loans are minimised, which effectively reduces the probability of bank insolvency. Rising high rates of inflation, on the other hand, is associated with relatively high and increasing costs of production, which in turn reduce the borrowers' ability to repay their loans. Consequently, the banks accumulate large amounts of non-performing loans, which increase the probability of bank insolvency. Depositors respond to the news that banks are accumulating large amounts of non-performing loans by rushing in a panic to withdraw their deposits in full, forcing the banks to liquidate their assets at a loss, which effectively increases the probability of bank insolvency. It has also been argued that high inflation rates cause uncertainty among economic agents, compelling them to protect themselves from the rising price instability, which leads to a slow-down in economic activity and hence an impairment of their ability to repay their loans (see for example, Li, 2006; Ghosh & Phillips, 1998). As non-performing loans start to accumulate, the probability of bank insolvency also increases.

Tables 7.5 and 7.6 also show that interest rates and the probability of bank insolvency, narrowly and broadly defined, are positively correlated. A rise in interest rates reflects an increase in the cost of loans and loss of income (associated with the higher cost of debt service), which

Table 7.5 Regression results of banking instability measured by the insolvency problem of banks broadly defined

Dependent variable: Insolvency broadly defined			
Variable	Coefficient	Standard Error	P-Value
GDP growth	−0.4249*	0.2360	0.0720
CPI inflation	−0.6857***	0.1987	0.0010
CPI inflation squared	0.0143***	0.0045	0.0020
Interest rates	1.3512***	0.4685	0.0040
Change in exchange rates	30.0356***	7.2271	0.0000
M2 as a ratio of foreign exchange reserves	0.1029	0.9000	0.9090
Political stability	−0.6060*	3.2050	0.0590
Rule of law	−1.6084	4.7139	0.7330
Number of observations = 85			
Wald chi2 (8) = 40.50			
Prob > chi2 = 0.0000			

NB Test critical values: *,**,*** represent 10%, 5% and 1%, respectively
Source Computed by the authors

Table 7.6 Regression results of banking instability measured by the insolvency problem of banks narrowly defined

Dependent variable: Insolvency narrowly defined			
Variable	Coefficient	Standard Error	P-Value
GDP growth	−0.4644*	0.2508	0.0640
CPI inflation	−0.7138***	0.2297	0.0020
CPI inflation squared	0.0150***	0.0051	0.0030
Interest rates	1.3963***	0.5289	0.0080
Change in exchange rates	30.8288***	7.6582	0.0000
M2 as a ratio of foreign exchange reserves	0.1852	0.8898	0.8350
Political stability	−6.4878*	3.7334	0.0820
Rule of law	−2.7866	5.1881	0.5910
Number of observations = 85			
Wald chi2 (8) = 37.47			
Prob > chi2 = 0.0000			

NB Test critical values: *,**,*** represent 10%, 5% and 1%, respectively
Source Computed by the authors

adversely affects borrowers' ability to repay their loans, leading to an accumulation of large amounts of non-performing loans and hence an increase in the probability of bank insolvency. On their part, banks view an increase in interest rates as a rise in the cost of funds (e.g. deposits). The high cost of funds coupled with accumulating large amounts of non-performing loans increase the banks' probability of bank insolvency. Therefore, rising interest rates tend to increase the probability of bank insolvency (both narrowly and broadly defined), whether viewed from the banks' or borrowers' viewpoint.

The study employs the exchange rate of the Ghanaian Cedi against the US dollar and the ratio of aggregate money supply to foreign exchange reserves to capture banks' vulnerability to sudden capital outflows triggered by the banks' exposure to foreign exchange risk. Tables 7.5 and 7.6 show that a depreciation of the Ghanaian Cedi against the US dollar increases the probability of bank insolvency. A depreciation of the Ghanaian Cedi against the US dollar is expected to trigger withdrawals of domestic currency deposits to purchase foreign assets. The reverse also holds. The larger the extent to which the Ghanaian Cedi depreciates against the US dollar, the higher the vulnerability of the domestic banks to foreign exchange risk, and consequently, the higher the probability of bank insolvency, both narrowly and broadly defined.

The ratio of aggregate money supply to foreign exchange reserves is also observed to be positively related to the probability of bank insolvency, narrowly or broadly defined. This may be explained by the behaviour of investors to shift their portfolios between domestic and foreign assets depending on the investment climate. It is highly likely that an unfavourable investment climate triggers capital outflows, leading to a decline of the ratio of M2 to foreign exchange reserves and consequently an increase in the probability of bank insolvency, narrowly or broadly defined.

Two variables are used to capture the quality of the regulatory and institutional environment in Ghana and their impact on banking instability. Tables 7.5 and 7.6 show no evidence of a relationship between banking instability characterised by bank insolvency, both narrowly and broadly defined, and the rule of law. This finding contradicts the earlier finding reported in Tables 7.3 and 7.4, that the rule of law plays an important role in explaining banking instability (characterised by bank runs, both narrowly and broadly defined). This connotes that people's "perceptions of the extent to which agents have confidence in and abide by the rules of society, and in particular the quality of contract enforcement, property rights, the police and the courts, as well as the likelihood of crime and violence" (see Kaufmann et al., 2010) do not affect bank insolvency in Ghana. It is, however, observed in Tables 5 and 6 that political stability (and the absence of violence/terrorism) significantly affects bank insolvency in the country, both narrowly and broadly defined. The tables show that an increase in the "likelihood of political instability and/or politically-motivated violence, including terrorism" (see Kaufmann et al., 2010) increases the probability of banking instability, narrowly and broadly defined.

7.5 Summary, Conclusions and Policy Recommendations

This chapter sets out to investigate macroeconomic factors that explain banking instability in Ghana. We categorise macroeconomic indicators into three groups, namely, macroeconomic factors that are expected to affect the quality of bank assets (consumer price inflation, real GDP growth and interest rates); factors that are expected to capture the exposure of local banks to foreign exchange risk (exchange rate depreciation/appreciation and the ratio of M2 to foreign exchange reserves); and factors that capture the quality of regulatory and institutional structures in the country (political stability and the rule of law).

Among the macroeconomic factors that are expected to affect the quality of bank assets, we find that an increase in real output reduces the probability of bank instability, whether characterised by bank runs or the insolvency problem of banks. We also establish that inflation rates have a curvilinear relationship with both measures of banking instability. At low and moderate levels of inflation, the probability of banking instability and inflation rates are inversely related, while at high levels of inflation, accelerating rates of inflation tend to increase the probability of banking instability. The study also finds that a rise in interest rates increases the probability of banking instability measured by both bank runs and insolvency of banks, narrowly and broadly defined. Among the two factors that are expected to capture the exposure of local banks to foreign exchange risk, the study establishes that a depreciation of the Ghanaian Cedi vis-à-vis the US dollar and a rise in aggregate money supply as a ratio of foreign exchange reserves increase the probability of banking instability.

There are two variables that are used to capture the quality of the regulatory and institutional environment in Ghana and its impact on the probability of banking instability. These are political stability and the rule of law. The study finds no evidence of a causal relationship between political stability and the probability of banking instability measured by bank runs, both narrowly and broadly defined. It is, however, established that improvements in political stability reduce the probability of banking instability characterised by the insolvency problem of banks. The findings are consistent in both the narrow and broad measures of bank insolvency.

The study also finds no evidence of a causal relationship between the rule of law and the probability of banking instability measured by bank insolvency in Ghana. However, we find that improvements in the rule of law reduce the probability of banking instability measured by bank runs.

Against the foregoing, it is recommended that Ghana must formulate policies that stimulate the growth of real output, keep interest rates low and build foreign exchange reserves (relative to aggregate money supply). There is also need for a further study to establish the level of inflation that aligns to optimal banking stability in Ghana. The optimal inflation threshold will guide policymakers in Ghana on the rate of inflation below which increasing inflation reduces the probability of banking instability and beyond which increasing inflation increases the probability of banking instability.

Following the finding that a depreciation/an appreciation of the Ghanaian Cedi against the US dollar increases/decreases (assuming symmetric responses) banking instability in Ghana, we recommend that the authorities should formulate policies that stabilise the exchange rate of the Ghanaian Cedi vis-à-vis major trading currencies. In this way, the domestic currency will not be propelling banking instability without compromising with the effects of a depreciation/an appreciation of the Cedi on other sectors of the economy. Finally, we recommend that Ghana should aim at improving its governance machinery to achieve political stability and strengthen the rule of law. As the study has indicated, strengthening the rule of law reduces the probability of banking instability measured by bank runs while increasing political stability reduces the probability of bank runs measured by the insolvency problem of banks.

The primary limitation of the study was the unavailability of data in the required frequency. Among the macroeconomic indicators, inflation, interest rates and exchange rates were available in quarterly frequency while real GDP growth, the ratio of M2 to foreign exchange reserves, political stability and the rule of law were not. To ensure that all the data have the same frequency, we interpolated real GDP growth, the ratio of M2 to foreign exchange reserves, political stability and the rule of law from annual to quarterly frequency.

Appendix I

Table A7.1 List of registered banks in Ghana

No.	*Bank Name*
1	Absa Bank Ghana Limited
2	Access Bank (Ghana) Plc
3	Agricultural Development Bank Limited
4	Bank of Africa Ghana Limited
5	CalBank PLC
6	Consolidated Bank Ghana Limited
7	Ecobank Ghana Limited
8	FBNBank (Ghana) Limited
9	Fidelity Bank Ghana Limited
10	First Atlantic Bank Limited
11	First National Bank (Ghana) Limited
12	GCB Bank Limited
13	Guaranty Trust Bank (Ghana) Limited
14	National Investment Bank Limited
15	OmniBSIC Bank Ghana Limited
16	Prudential Bank Limited
17	Republic Bank (Ghana) PLC
18	Societe General (Ghana) Limited
19	Stanbic Bank Ghana Limited
20	Standard Chartered Bank (Ghana) Limited
21	United Bank for Africa (Ghana) Limited
22	Universal Merchant Bank Limited
23	Zenith Bank (Ghana) Limited

Source Bank of Ghana website https://www.bog.gov.gh/supervision-regulation/registered-institutions/banks/

References

Adao, B., & Temzelides, T. (1998). Sequential equilibrium and competition in a Diamond-Dybvig banking model. *Journal of Economic Dynamics, 1*, 859–877.

Adusei, M. (2015). The impact of bank size and funding risk on bank stability. *Cogent Economics & Finance, 3*(1), 1111489.

Aithnard, P., Bodo, G., Mohammed, S. and Zindogo, S. (2013). Banking in central and West Africa: A growth story. *Banking in sub-Saharan Africa*, 7.

Almeida, A., Fry, M.J., Goodhart, C., 2004. *Central banking in developing countries: Objectives, activities and independence* Routledge.

Antwi, J. (2020). *Bank reforms, competition, and stability in the Ghana banking industry* (Doctoral dissertation, University of Delaware).
Allen, F., & Gale, D. (1998). Optimal financial crises. *Journal of Finance, 53*(4), 1245–1284.
Bank of Ghana (2022). Bank of Ghana website, Accessed on 26 January 2022. https://www.bog.gov.gh/
Bank of Ghana (2017). GCB Bank takes over UT Bank limited and capital limited, www.bog.gov.gh
Bank of Ghana (2018). Government established new indigenous bank; Bank of Ghana revokes licences of five banks and appoints receiver in respect of their assets and liabilities, Bank of Uganda Press Statement accessed on 02 June 2021. Retrieved at https://www.bog.gov.gh/wp-content/uploads/2019/07/PRESS-RELEASE-Grand-Final-August-2018.pdf
Brook, A.-M., Karagedikli, Ö., & Scrimgeour, D. (2002). An optimal inflation target for New Zealand: Lessons from the literature. *Reserve Bank of New Zealand Bulletin, 65*, 5–16.
Caprio, G., and Klingebiel, D. (1997). Bank insolvency: Bad luck, bad policy or bad banking? *Annual World Bank conference on development economics*, 1996.
Carmona, G. (2004). On the existence of equilibrium bank runs in a Diamond-Dybvig environment, Unpublished Manuscript, Universidade Nova de Lisboa.
Chari, V., & Jagannathan, R. (1988). Banking panics, Information, and rational expectations equilibrium. *Journal of Finance, 43*(3), 749–761.
Chartered Institute of Bankers, Ghana (2021). Corporate members in good standing 2019. Retrieved on 07 June 2021. Accessed on https://www.cibgh.org/membership/corporate-members-good-standing
Demirgüç-Kunt, A., & Detragiache, E. (2002). Does deposit insurance increase banking system instability? An empirical investigation. *Journal of Monetary Economics, 49*, 1373–1406.
Diamond, D., & Dybvig, P. (1983). Bank runs, deposit insurance and liquidity. *Journal of Political Economy, 91*(3), 401–419.
Drigă, I., and Dura, C. (2014). The financial sector and the role of banks in economic development. In *6th International Multidisciplinary Symposium Universitaria SIMPRO* (pp. 10–11).
Eichengreen, B., Rose, A., & Wyplosz, C. (1995). Exchange market mayhem: The antecedents and aftermath of speculative attacks. *Economic Policy, 21*, 249–296.
Eichengreen, B., Rose, A., & Wyplosz, C. (1996a). Contagious currency crises: First tests. *Scandinavian Journal of Economics, 98*(4), 463–484.
Eichengreen, B., Rose, A. and Wyplosz, C. (1996b). Speculative attacks on pegged exchange rates: An empirical exploration with special reference to the European monetary system. In M.B. Canzoneri, W.J. Ethier, and V. Grilli,

(Eds.), *The New Transatlantic Economy*: 1996b, Cambridge, Cambridge University Press.

Fisher, I. (1911). *The Purchasing power of money: Its determination and relation to credit, interest and money*. MacMillan.

Foley-Fisher, N., Narajabad, B., & Verani, S. (2020). Self-fulfilling runs: Evidence from the US Life Insurance Industry. *Journal of Political Economy, 128*(9), 3520–3569.

Ghosh, A., & Phillips, S. (1998). Warning: Inflation may be harmful to your growth. *IMF Staff Papers, 45*, 672.

Global, I. H. S. (2017). Eviews 10 user's guide I. *IHS Global Inc.*Grilli, L., and Rampichini, C. (2014). Ordered logit model. *Encyclopedia of quality of life and well-being research*, 4510–4513.

Grevenitis, K., Sakellariou, I., & Kefalas, P. (2020). Emotional Agents Make a (Bank) Run. In N. Bassiliades, G. Chalkiadakis, D. de Jonge (Eds.), *Multi-Agent Systems and Agreement Technologies*. EUMAS AT 2020 2020. Lecture Notes in Computer Science, vol 12520. Springer, Cham: Switzerland. https://doi.org/10.1007/978-3-030-66412-1_12.

Gujarati, D. (2003). *Basic econometrics*. The McGraw Hill Companies Inc.

International Monetary Fund (2011). Ghana: Financial system stability assessment update, International Monetary Fund Headquarters: Washington DC. Accessed on 21 June 2021. Retrieved on https://www.imf.org/en/Publications/CR/Issues/2016/12/31/Ghana-Financial-System-Stability-Assessment-Update-24917

Ishii, S., Habermeier, K., Canales-Kriljenko, J. I., Laurens, B., Leimone, J., & Vadasz, J. (2002). Capital account liberalization and financial sector stability. IMF Occasional Paper 211, International Monetary Fund, Washington, USA.

Jacobs, J. P., & Kuper, G. H. (2007). *Currency crises in Asia: A multivariate logit approach*. Integration, innovation and challenges. Emerald Group Publishing Limited.

Kaufmann, D., Kraay, A. and Mastruzzi, M. (2010). "The Worldwide Governance Indicators: Methodology and Analytical Issues". *World Bank Policy Research Working Paper No. 5430* (http://papers.ssrn.com/sol3/papers.cfm?abstract_id=1682130, accessed on 01 December 2021).

Kamasa, K., Mochiah, I., Doku, A. K., & Forson, P. (2020). The impact of financial sector reforms on foreign direct investment in an emerging economy; Empirical evidence from Ghana. *Journal of Humanities and Applied Social Sciences, 2*(4), 271–284.

Kong, Y., Atuahene, S. A., Bentum-Micah, G., & Agyapong, M. K. (2020). Corporate governance and banking stability: The Case of Universal Banks in Ghana. *International Journal of Economics & Business Administration (IJEBA), 8*(1), 325–352.

Kwakye, J. K. (2012). Financial intermediation and the cost of credit in Ghana.

Leiva, D. R., & Mendizábal, H. R. (2019). Self-fulfilling runs and endogenous liquidity creation. *Journal of Financial Stability, 45*, 100704.

Li, M., 2006. *Inflation and economic growth, department of economics*. University of Alberta.

Mertens, A., & Urga, G. (2001). Efficiency, scale and scope economies in the Ukrainian banking sector in 1998. *Emerging Markets Review, 2*(3), 292–308.

Morar, D. (2011). *Inflation threshold and nonlinearity: Implications for inflation targeting in South Africa, department of economics and economic history*. Rhodes University.

Moyo, J., Nandwa, B., Council, D. E., Oduor, J., and Simpasa, A. (2014). Financial sector reforms, competition and banking system stability in Sub-Saharan Africa. *New perspectives*.

Ngalawa, H. (2012). Banking instability and deposit insurance in low income countries. *Journal for Studies in Economics and Econometrics, 36*(3), 1–24.

Ngalawa, H., Tchanatchana, F., & Viegi, N. (2016). Banking instability and deposit insurance: The role of moral hazard. *Journal of Applied Economics, XIX, 2*, 323–350.

Ngari, J.M.K. and Muiruri, J.K., 2014. Effects of financial innovations on the financial performance of commercial banks in Kenya.

Obuobi, B., Nketiah, E., Awuah, F., & Amadi, A. G. (2019). Recapitalization of banks: Analysis of the Ghana banking industry. *Open Journal of Business and Management, 8*(01), 78–103.

PwC (2021). Digital currencies: Is the Ghanaian banking sector ready? 2021 PwC Ghana banking survey report, accessed on 26 January 2021. https://www.pwc.com/gh/en/assets/pdf/ghana-banking-survey-report-2021.pdf

Von Hagen, J., & Ho, T. K. (2007). Money market pressure and the determinants of banking crises. *Journal of Money, Credit and Banking, 39*(5), 1037–1066.

Yensu, J., Yusif, H., Tetteh, E., Asumadu, G., & Atuilik, D. A. (2021). Main determinants of banks' stability: Evidence from commercial banks in Ghana. *Journal of Finance and Economics, 9*(2), 42–52.

CHAPTER 8

Financial Dualism in Ghana: The Implications for Monetary Policy on Loan Variations

P. Adjei and R. K. Ayisi

8.1 Introduction

The financial sector plays critical role in monetary policy. It is the medium by which monetary actions influence the real sector (Ehrmann et al., 2003; Kashyap & Stein, 1997). The central bank can influence the banks because of their regulatory oversight.[1] This notwithstanding, Ghana's financial sector (like many other developing countries) consists of two

[1] The central bank regulates the banks using various framework such as the mandatory reserves, discount windows among others.

P. Adjei (✉) · R. K. Ayisi
Department of Economics, University of Ghana, Legon, Ghana
e-mail: pradjei@ug.edu.gh

R. K. Ayisi
e-mail: rayisi@ug.edu.gh

J. A. Peprah et al. (eds.), *Financial Sector Development in Ghana*, Palgrave Macmillan Studies in Banking and Financial Institutions,
https://doi.org/10.1007/978-3-031-09345-6_8

types of banking activities; the traditional universal banking and micro-financing which associates differently within the regulatory framework with the central bank.[2]

Informal[3] financial activities have developed into important fabric of the development process of various economies. They complement the traditional institutions by mopping up extra liquidity (savings) in the economy which otherwise may be unreached by the traditional sector. In addition, the informal sector provides social and financial intermediation services to about 80% of developing economies that mainly consist of informal activities and low-income earners.

Although, the growth in size and volume of informal finance is beneficial in the growth-finance nexus literature, the increase in informal credit activities raises some concerns. Anecdotal evidence firstly suggests that there is overlap in operations between informal and universal banks in most developing economies. While informal finance emerged complementary to cater for segment of the credit markets not captured by the formal banks, there is evidence of formal and informal banks functioning in the same market in recent times. This has caused the informal banks to compete with the formal banks instead of complementing each other in the credit market with consequential implication on their strategic interactions, which may generate contagion risk for the entire financial ecosystem and monetary policy effectiveness.

The literature shows that monetary policy effect on small firms and borrowers is immersed than it has on large ones (Gertler & Gilchrist, 1993; Foley-Fisher et. al., 2016). Also, Kashyap et al. (1992) show that monetary policy changes the credit mix of economies with negative effect on borrowers and smaller firms. This provides microfinance institutions the opportunity to fill-in since their characteristics insulate them of the pressures the formal financial institutions face. The study explored the validity or otherwise of this.

[2] The 2014 Financial Stability Board report by the IMF estimated the size of world financial system captured by non-bank financial intermediation is about 40%.

[3] For this study, we classified banks i.e., deposit taking institutions into formal banks and informal banks. The term "formal bank" used interchangeably with traditional or universal banks refers to deposit taking institutions that are mandated to hold reserves with the central bank i.e. operates under the required reserve regulations of the central bank. The term "informal bank" used interchangeably with microfinance consists of deposit taking institutions that are not obliged to keep reserve with the central bank.

In Ghana, the informal financial institutions are not under the reserve regulation of the central bank,[4] and as such, monetary actions do not affect their lending directly. For a small informal organisation, a monetary policy can be taken as an exogenous shock. Following this, the study examines the case whether informal financial organisations differ from their formal counterparts in their response to monetary policy, and if so, whether their loan portfolio mix changes exactly as the formal bank or there are differences?

We explored these questions using micro data of banks to identify the changes in loan supply of both formal banks and informal banks following monetary shocks. The study established that variations in credit supplies of traditional banks are significantly influenced by the size of the bank, the capitalization, and policy rate. While the size and liquidity position passively influence loan variations, the capital structure of the bank and policy rate negatively influence loan variations. The study further revealed that monetary policy is insignificant in determining variations in credits of informal banks.

The rest of the paper is organised as follows. Section 8.2 presents the literature review which also include a structure of the financial service industry in Ghana. Section 8.3 presents the methodology used for the empirical investigation. Section 8.4 presents the results and discussion. Section 8.5 concludes and provides some policy recommendations.

8.2 Literature Review

8.2.1 Financial Dualism and the Design of Monetary Policy

Financial dualism[5] occurs when a formal financial sector coexists with an informal financial sector in an economy. Aryeetey (2008), Guirkinger (2008), and Madestam (2014) explained financial dualism as the situation in which formal or advanced and informal or traditional monetary subdivisions of a particular economy co-occur or exist simultaneously

[4] The microfinance institutions operate under the non-bank financial Institution Act 2008 (Act 774), which does not allow them to operate under the required reserve regulations.

[5] The term financial dualism was introduced by Professor Hla Myint in his attempt to explain the gap between the formal and informal financial sector in terms of interest rate but credited to Julius Herman Boeke who analysed the financial sector of the colonial Indonesian economy in the mid-1900s.

with price differentials. According to Myint (1985), the formal financial sector is mainly characterised by a highly specialised or supervised working environment, capital-intensive technology, and a low and fixed interest rate that mostly provides services to commercial and retail institutions. Institutions within this sector include banks, insurance, and investment companies, real estate firms, among others. The informal financial sector on the other hand is dominated by individuals' institutions that are self-reliant and have no financial support from the government hence interest placed on loans is comparably higher. Hoff and Stiglitz (1990), Aryeetey and Gockel (1991), and Guirkinger (2008) remarked that traditional monetary systems usually exhibit steep lending rates as a result of small credit facilities against the backdrop of the higher rates of default, whereas the advanced monetary segment offers relatively cheaper costs of borrowing to limited consumers.

In a simplified business cycle, the financial sector is seen to play a crucial role in the exchange of goods and services. The sector is charged with handling the buying and selling of financial instruments through savings and investments. They receive savings from the household, firms, and government and sell investment packages to them in return. The government borrows and sells to these financial institutions as well. While most household-financial sector trade is done through the informal financial sector, the financial trade activities of firms and government are undertaken through the formal financial sector. A study by Aryeetey (1995), showed that about 90% of households took loans from friends, relatives, or landlords in the late 1900s. Recently, the IMF Financial Stability Board in its 2014 report on Global shadow banking monitoring estimates that informal credit constitutes about 40% of the total credit intermediation in the world.

Shem and Atieno (2001) share the view that dualism in low-income countries is usually as a response to the deficiencies of the formal financial system and excessive regulation or the dualism of their economies. If dualism is triggered due to excessive regulations, then it may be as a result of policies such as interest rate controls, confiscatory reserve requirements, overvaluation of the domestic currency, and credit subsidies. Financial dualism on account of dualism of the economies may be due to illiteracy, transportation and communication network, inequitable distribution of social services, imperfect information, transaction cost, collateralisation, and absence of or weak insurance schemes.

Aryeetey (2005) and Guirkinger (2008) propounded that segmentation persists as a very troubling consequence of financial dualism. This is because the variation of the two systems allows for financial services to be made available to diverse groups of clienteles, in accordance with the peculiarity of their financial service outlets. This potentially could create roadblocks that impede monetary substitution between the formal and informal systems, which subsequently constricts the option of consumers when accessing financial services. This is often associated with exclusion implications, where deficit units incapable of satisfying requirements in the formal sector can't secure financing, and at the same time will be unable to afford the steep lending rates common to informal systems, which altogether constrains the range of financial activity of deficit units or households. Moreover, segmentation can also cause more adverse effects in terms of equity challenges, since the two systems cater for diverse socio-economic household and business units, thus the underprivileged who live off the informal sector will become worse off with regards to the investment returns.

Asiamah (2017) reported that the connection between advanced and traditional financial intermediaries is a crucial matter of consideration for tackling the adverse impacts of financial dualism, noting that the classes of players can operate as either substitutes or complements. Levenson and Besley (1996), and Besley and Levenson (1996) explained substitution as the situation in which formal financial outlets are non-existent or weakly developed thus precipitating demand for alternatives solutions from traditional players. Supporters of substitution maintain that inadequate or insufficient services from formal outlets will not necessarily be an absolute barrier to the progress of traditional institutions since they can manoeuvre to fill those gaps to stimulate growth. Concerning complementary association, Morduch (1995) and Stiglitz (1989) observed that informal players operate as safety-nets to cover the market deficits usually associated with mainstream agents. Simply, traditional players or institutions function as a support system to their formal contemporaries. In the same vein, deficit units or borrowers across both systems are able to safeguard themselves from deficiencies in the traditional sector by relying on formal players, as Asiamah (2017) stated that in numerous developing economies where households access credit from both traditional and advanced intermediaries, the nature of relationship is all the more complementary.

Most financial services in African countries are being handled by the informal financial sector. In Ghana, the existence of the informal financial sector can be traced back to years before the British colonial rule introduced the formal financial sector. A typical example of such is the "Susu" system of saving money. Work by Adeusi (2012) on financial dualism in Nigeria showed that despite monetary efforts by the Central bank to reduce the size of the informal sector, the sector keeps increasing. Okurut et al (2004) analysed the reason behind Uganda's rise in the informal financial sector regardless of the development of the formal sector, it was established that gender and the assets of the household have led to the increase in the demand for the informal sector. Other studies have attributed the increase in demand for the informal financial sector in less developed countries on the level of education of the family head, size of the family, the number of dependents, credit history, and lack of access to the formal financial sector services, among others.

Financial dualism nevertheless has various effects on the financial market and the economy (see Igoni et al., 2020; Hung, 2015). The primary effect is via the different interest rates charged by the sectors. The informal financial sector charges a higher interest rate on loans compared to the formal sector. This is because of the indirect monopoly power created by moneylenders, landlords, and traders in the informal sector and the ex-post moral hazard that is usually high among informal borrowers. Borrowers, therefore, have no option other than to go for these services. Also, because access to credit is simple and does not require any legal documentation, the risk premium is higher hence a higher interest rate. According to data from the Central Bank of China, in Taiwan between the periods of 1970–84 the average interest rate in the informal sector was about 24.15% per annum and that of the formal sector at 12.38% per annum, showing a higher rate in the informal financial sector. Similar evidence is documented for Ghana. While the average lending rate among the formal banks is about 25%, that of the informal stands about 60% per annum.

Liang (1998) postulated that in a country where financial dualism exists, the determination of interest rates is a critical and very sensitive monetary policy mechanism. Monetary agents in developing nations, drawing inspiration from Keynesian economics, dispose to keep interest rates at artificially low levels, with the premise of accentuating investment. However, this type of monetary regime has increasingly come under severe criticism in recent years. Earlier studies such as argued that policy rates should be raised because the policy of small interest rates leads to unbridled credit rationing and inefficient appropriation of funds, not discounting the primary reason for inflationary troubles in most developing economies is due to low policy rates. Liang (1998) comments that while there is some truth to Tsiang's assertion, his assessment is only "a partial equilibrium" proposition. Though this issue has featured prominently in policy discussion recently, the literature has been salient.

Through the intermediary role of banks, Central Banks influence the real sector by manipulating the amount of money in the economy. Central banks control the flow of money in the system using instruments such as the interest rate, Open Market Operations, required reserves, and discount rate. Within the current Inflation Targeting regime practised in Ghana, the interest rate is the main channel of monetary transmission.

As discussed earlier, the interest rate in a financial dualism economy differs. The policy/discount rate set by the central bank influences the retail rate used by the formal financial sector such as banks, investment institutions, and insurance institutions, among others. A higher discount rate will result in a higher cost of borrowing for banks hence an increased in the interest rate charged by banks. A lower interest rate in the formal financial sector has led to a reduction in saving levels and an increase in the demand for loans. This will in effect harm the informal sector as the supply of capital in the sector reduces due to the higher rates charged in the sector. In the long run, it will be observed that monetary transactions in the economy will reduce to favour non-monetary transactions. Carpenter (1999) studied the linkage between the informal financial sector and monetary policy, he theorised the following; an expansionary monetary policy leads to an increase in the interest rate, therefore, we expect a higher interest rate in the formal financial sector but not higher than that of the informal. His analysis suggested that the formal and informal financial sector are substitutes hence there is an expectation of price fall in the informal sector, but on average, an increase in formal sector interest rate will create output expansion and excess demand for credit, leading to a much higher interest rate in the informal sector. A study by Ghosh and Kumar (2015), showed that monetary contractionary policies that lead to an increase in the cost of borrowing from the formal sector will also have a similar effect in the informal sector given the formal-informal inter-relationship, on average, the informal financial sector is expected to have a reduction in the level of borrowing because of the cost associated with borrowing of credit that outweighs the benefits.

He also found out that, the reduction in borrowing in the informal sector is only affected by moneylenders but landlords, friends, and relatives' level of borrowing increases in the long run after a contractionary policy has been implemented.

These issues have led to slower growth in the financial market in most LDCs. To strengthen the financial sector, it is vital to integrate both the formal and informal financial sector, this can be done by providing financial services to rural areas and ensuring these services are in the interest of the people. Monetary policy Committee should capture the role of the informal financial sector in its decision-making as it has been established that the policies implemented affected the informal sector through the formal sector.

8.2.2 *Brief Empirical Literature: Bank Lending Channel*

The main channels of monetary policy transmission are the interest rate, credit, exchange rate, and other asset price channels. The credit channel, which is the focus of this study has two mechanism—the bank-lending channel (BLC) and balance-sheet channel (BSC) (Bernanke & Gertler, 1995). The BSC emphasises the effect monetary policy change on the balance sheet of borrowers, while the BLC emphasises the impact monetary policy action has on the credit supply of bank.

A channel of monetary transmission that emphasises the role of banks in transmitting monetary impulses is the bank lending channel, which posits how changes in monetary policy shift banks' credit supply schedules. The empirical literature on bank lending channel is well documented for developed countries. Most of these studies use disaggregated bank-level data (see Altunvas et al., 2002; Angeloni et al., 2003; Ashcraft, 2006; Ehrmann et al., 2003; Gambacorta, 2005; Huang, 2003; Kashyap et al., 1995; Kishan & Opiela, 2000; Walsh, 2003). The general conclusion in most of these studies shows that monetary policy tightening causes bank credit (loans) to decrease, resulting in a negative impact on the economy. While empirical studies (see Aysan et al., 2018; Kashyap & Stein, 2007; Kishan & Opiela, 2000; Oliner & Rudebusch, 1996; Pruteanu-Podpiera, 2007) have explored this channel mostly for the traditional banking system, little or no evidence is found for microfinance or the informal financial institutions, unfortunately.

Peek and Rosengren (1995) analysis on bank lending and its relation to the monetary policy stated that the hitherto monetary transmission view ("the money view") which focuses on reserve requirement constraint on the banks has been augmented due to changes in the supply of bank credits to "the lending view" which concentrates on capital constraint as a way of examining the magnitude of the effect of monetary policy over time. Their study showed the importance of capital constraint on bank lending with consequent effect on the lending channel of monetary.

Kashyap and Stein (1995) in their attempt to define the lending view, pointed out the difference between "the money view" and "the lending view". The authors stated that the difference between the two views is that, while the latter perceives the financial system as a three-asset world (money, bonds, and bank loans), the former observes it as a two-asset world (money and bonds). They again pointed out that bank loans growth in small commercial banks was responsive to monetary policy changes. Although their data did not support these findings, they believed that lending channels could be reduced when banks are more capital oriented.

Pruteanu-Podpiera (2007) opines that the characteristics of the market supply curve of credits allow for operationalizing the credit channel of the monetary transmission mechanism. Specifically, the credit channel stresses that monetary policy affects loan demand directly through both the interest rate channel and the supply of loans. This peculiarity makes the credit channel different from the depictions made by the interest rate channel. By extension, economies with more borrowers depending on banks have the credit channel to be very important transmission mechanism. This is due to the experience of the existence of financial frictions that might significantly affect the monetary transmission mechanism.

According to Matousek and Sarantis (2009), the Bank Lending Channel is a combination of lending binding constraint and a deposit market constraint where the bank with fragile balance sheets becomes more sensitive to monetary shocks compared to a bank with strong balance sheets. To measure how the strength of a bank's balance sheets could affect the response of bank loans to a change in monetary policy, previous works of literature such as (Kashyap & Stein, 1995, 2000; Kishan & Opiela, 2000; Peek & Rosengren, 1995) pointed out three main characteristics; asset size, bank capitalisation, and liquidity.

In analysing the liquidity nature of banks and their relation to their monetary policy, Kashyap and Stein's (2000) study revealed that the more liquid a bank is, the more responsive it is to monetary policy. Contrary to Stein's findings, Ehrmann et al. (2003) analysis showed the inverse where less liquid banks responded strongly to changes in monetary policy. Concerning asset size, Ashcraft (2006), revealed that aggregate lending in state-owned banking markets is more responsive to monetary policy compared to other banks. They also showed that the importance that small firms have placed on bank loans is not enough for the lending channel to play a significant role in how monetary policy works.

Kishan and Opiela's (2000) study on both bank size and capital link to the bank lending channel and monetary policy grouped bank capital into three categories; undercapitalised banks, well-capitalised and adequately capitalised. For small banks, they showed that the growth of these banks under all three categories was significantly affected by monetary policy. Altunbaş et al. (2002) also showed that irrespective of the bank size, underutilised banks responded more to changes in policy. Matousek and Sarantis (2009) study for six Central and Eastern European Countries, showed that overall, banks size and liquidity were more significant in determining the responsiveness of banks to monetary policy. For Portugal however, Favero et al. (1999) noted that the importance of bank lending channels is great when there is a less-capitalised bank.

During the monetary restriction in 1992, Favero et al. (1999) showed that for the four European countries they studied; Germany, France, Italy, and Spain, there was no evidence of how responsive bank loans are to monetary policy, however, in cases where short-term interest rate is used as a proxy for monetary policy, Debondt (2013) study showed that bank lending channel was present in Germany. The author later replaced the short-term interest rate with monetary condition index as a proxy for monetary policy and found the existence of a bank lending channel in France and Italy. In the case of the USA, Oliner and Rudebusch (1995) found no evidence in support of bank lending channels.

For African countries, Sichei (2005) showed the presence of bank lending channel in South Africa using bank size and capital-asset ratio. The study pointed out two conditions for bank lending channels to work in South Africa; there should be bank-dependent customers in South Africa which holds in the country and monetary policy by the central bank should affect the supply of loans which the study seeks to examine. The paper established that contractionary monetary policy affects mostly smaller banks and their customers with financial innovation and bank regulations affecting the effectiveness of the monetary policy.

Analysis for 31 developing countries done by Sanfilippo-Azofra et al. (2018) showed that the supply of loans for banks in countries with less developed financial sector had no effect when monetary policy changes, however, for countries with the developed financial sector, bank lending channels were very effective and when their financial sector development increases, a negative effect between the monetary policy and bank lending channels becomes greater.

In examining the relationship between bank lending channel of monetary policy and its effect on the mortgage, Black et al. (2010), showed that bank lending channels were present only in transition banks where these banks respond to monetary contraction policy by reducing mortgage lending. However, for traditional banks and market-based banks, they did not find any evidence of bank lending channels.

In Ghana, Sarpong-Kumankuma et al. (2013) found a positive relationship between bank size and capital with bank lending behaviours. The author stated that bigger banks lend more compared to smaller banks. Their analysis showed that some macroeconomic indicators (exchange rate, Central bank Lending rate) had a negative impact on bank lending behaviours.

The empirical literature as showed above highlights the presence of bank lending channels in various scenarios. The literature further showed that the channel is influenced by factors such as capitalisation, sizes, and liquidity. These findings only support the traditional bank with no evidence for the informal banks and the possible effect from their interactions. This gap is the focus of this study. While enriching the literature from the Ghanaian perspective, we analyse the implications of the interactions between the traditional and informal banks on the lending transmission.

8.2.3 *Structure of the Financial Services Industry in Ghana*

To put the objective of this study in perspective, this section briefly reviews the financial landscape to show the interrelations between the formal and informal financial sector and the role it could play in their response to monetary shocks. The duality of the banking sector of Ghana is encapsulated in the distinction of "bank" and "non-bank" category that are required to operate in different credit market. Figure 8.1 gives the schematic representation of the financial system of Ghana.

From Fig. 8.1, the banking institutions are classified into either deposit taking or non-deposit taking institutions. The deposit taking institutions, in addition to advancing credit facilities, are allowed to accumulate

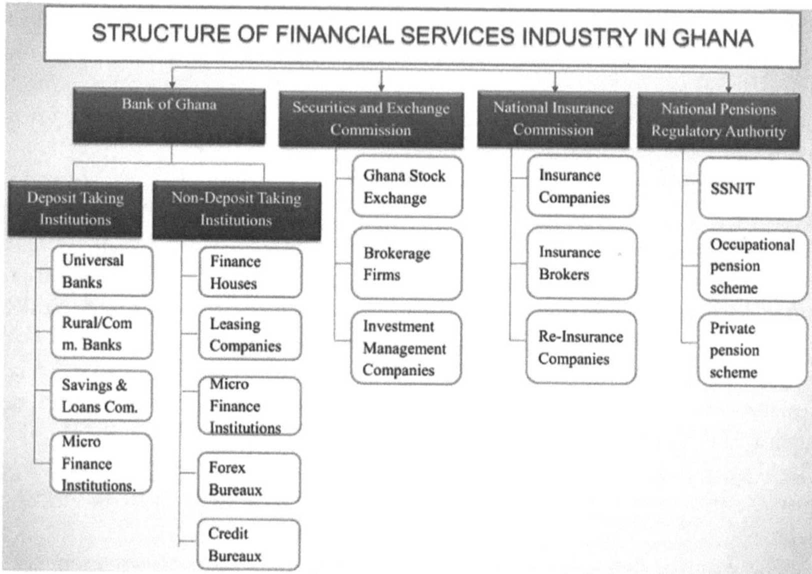

Fig. 8.1 Financial structure of Ghana (*Source* Bank of Ghana)

deposit. These categories of banks include the universal banks, rural and community banks (RCBs), non-bank financial institutions (NBFIs), savings and loans companies, and some microfinance institutions. The non-deposit taking institutions do not have the privilege of deposit collections but only advance credit facilities to borrowers. This group includes the finance houses, leasing companies, forex bureau, credit bureau, and some micro-finance.

The microfinance emerged complementary to cater for segment of the credit markets not captured by the traditional banks. Nonetheless, their activities overtime have grown to become competitive in the credit intermediation process. The scenario has resulted in financial dualism, a situation where universal banks coexist with the microfinance institutions under the same category of deposit taking but compete in the credit market rather than complementing each other.

Notwithstanding the parallel operations of these types of institutions, they operate under different regulatory environment. While the universal banks operate under banking of Ghana act, 2002 (Act 612) amended in

2016, the savings and loans, RCB, and microfinance are licensed under NBFI law. Depending on the type, they may be permitted to accumulate deposit or limited to loans. For example, the RCB and deposit taking microfinance offer banking services for both saving and loans, but the non-deposit institutions are only limited to loans. The differing regulatory environment renders informal credit activities not to fall strictly under the regulatory environment of central bank as the universal banks. While the universal banks are required by law to keep some reserve with the central bank, informal banks are not mandatory to do so. The implication is that informal institutions remain potentially to fill the gap created in the credit market as a result of the compositional change in credit mix due to tight monetary policy shock (Kashyap et al., 1992).

8.3 Methodology

8.3.1 Theoretical Framework

The bank lending channel is derived from the relationship between the cost of funds of bank and monetary policy shock via changes in the marginal cost of funding. Traditionally, the idea is premised on the ability of central banks to directly manipulate the level of deposits through the money multiplier mechanism (Bernanke & Blinder, 1988; Kashyap & Stein, 1995; Stein, 2012; Walsh, 2003). The idea relied on portfolio substitution arguments whereby policy tightening reduces the relative yields on deposits, inducing households to economise on them (Ehrmann et al., 2001; Kishan & Opiela, 2000). Following this, we posit in this study that given that both formal and informal credit are in the portfolio basket of household, a monetary shock will trigger a substitution of formal credit to informal credit given their rivalry in nature. In this regard, we show below a framework to espouse this assertion.

The starting point to analyse the channel is through optimisation on credit intermediation process.

As noted, both the formal and informal banks provide credit facility in the economy. As a result, the aggregate credit portfolio is a composite of formal and informal credit given as:

$$B = f\left(b^f, b^m\right) \tag{8.3.1}$$

where b represents credit and superscript f and m denote formal and informal respectively.

The formal bank under the required reserve regulation of the central bank engages in the production of deposit and loan services with a given technology represented by a cost function C.

Thus, maximises its profit subject to the loan demands considering management cost.

·Formally, subject to Eq. (8.3.1), the formal bank

$$Max\pi = \left\{r^f B^f - RB^f - C\right\} \tag{8.3.2}$$

The first order condition prescribes the loan reaction function specify in its compact form as.

$$B^f = f\left(r^f, R, \eta\right) \tag{8.3.3}$$

.

Where r^f denotes retail interest rate, R denotes the policy rate, and η captures all other variables (particularly bank-specific characteristics) that may influence credit reaction function.

We reparametrised Eq. (8.3.1) for both formal and informal banks. An increase in policy rate is expected to inversely affect credit change, thus a positive change in policy rate directly affect retail rates which consequently affect the credit demand inversely. This mechanism is, however, possible for the bank when its retail rate is tight to the policy rate via the interbank lending system, which is made possible via the central bank's role of lender of last resort through the required reserve regulation.

As indicated in Sect. 8.2.3, the classification of the two banks—formal and informal is along the line of the required reserve regulations. Therefore, the above holds for the formal bank but not for the informal bank. Following this, we expect policy rate to affect the credit demands of formal and informal bank differently.

8.3.2 Empirical Modelling

Broadly, the study explores how the variations in the loan supply of both formal banks and informal banks[6] are affected by policy instrument. To achieve this, based on the assumption that bank idiosyncratic characteristics determine the degree to which banks respond to monetary policy shocks, the study adopts a bank-level data to identify shifts in loan supply.[7]

Accordingly, we setup a panel regression model using Eq. (8.3.1) that specifies loan growth for each bank as a function of its lagged values, bank-specific characteristics (size, liquidity, and capitalisation), and policy variable (MPR). This draws on model from Bernanke and Blinder (1988). The specification is given as

$$\Delta L_{it} = \alpha_0 + \alpha_{00}\Delta L_{i,t-1} + \alpha_1 SIZE_{i,t} + \alpha_2 LIQ_{i,t} + \alpha_3 CAP_{i,t} + \alpha_4 MPR_t + \in_{It} \quad (8.3.4)$$

where ΔL is the percentage change in loans for bank i in the period from the previous; LIQ is the liquidity ratio of bank; CAP is the capitalisation of the bank.

To incorporate the objective of the study, we reparametrised based on Eq. (8.3.4) for the informal bank.

8.3.3 Data

The data is at bank level covering eleven (11) universal banks and five (5) microfinance institutions in Ghana. The universal banks selected included Ghana Commercial, Standard Chartered. Barclays, Cal, The Trust bank, International Commercial bank, First Atlantic, Ecobank, SG-SSB, PBL, and SBL. Due to acquisition and merger along the time frame used, the analysis resulted in unbalanced panel. The microfinance institutions selected are UTRAK, Opportunity, IZWE, BAYPORT, and Bonds. The

[6] Note that for this study the term "informal bank" is used loosely to represent formal microfinance institutions who are not under the required reserve framework of the central bank.

[7] Some strategy involves examining the reaction of bank loans to monetary policy shocks, using impulse response functions from a vector-autoregression (VAR) model. Though this setup does not allow the identification of supply and demand effects on credit growth, the study used it as a robustness check on the directions of credit change to policy shock and shock from the microfinance sector.

institutions were selected based on data availability. While universal banks were sampled between 2000 and 2017, the microfinance institutions were sampled from 2013 to 2019. The universal bank data is sourced from the Association of Bankers while that of microfinance source from their various balance sheets. For each bank, the balance sheet provides data on the total loan amount that is further split between short- and medium-term lending. The balance sheet data also allow us to construct the explanatory variables used in the analysis. The constructed explanatory variables included: bank size defined as the logarithm of total bank assets, the liquidity ratio of liquid holdings to total assets, capitalisation defined as the ratio of shareholders' equity to total assets. We, additionally, constructed dummy variables to capture ownership (state, foreign, private) and stock market participation (listed, unlisted). In addition to the bank characteristics variable, we included macroeconomic variables, which are gross domestic growth, GDP deflator, inflation, and growth in gross capital formation. The macro variables were sourced from the World Development Indicators (WDI).

8.4 Results and Discussions

We present the empirical estimates in Table 8.1. We estimated Eq. (8.3.4) using panel Estimated Generalised Least Square (EGLS) with cross-section weights estimator under the assumption of random variations across banks. We utilised this estimator due to the existence of cross-section dependence among the banks with potential of heteroskedasticity and serial correlations. Though the cross-section dependence may require more robust estimators such as CCEMG or AMG estimators developed respectively by Pesaran (2006) and Eberhardt and Teal (2010), these estimators apply the OLS for each panel members (after adjusting original model to a form that considers the cross-section dependence) and then take the average of the estimated coefficients to reach the panel coefficients, which requires more than 30 observations for each panel members. In our current model with less than 30 periods, we adopted the EGLS which is capable to address the embedded problem.

The estimated results are shown in Table 8.1. We estimate an extremely parsimonious model where loan variations are influenced only by bank-specific characteristics presented in panel (a). In panel (b), we include macroeconomic variables that may influence bank loan variations. Interestingly, the extremely parsimonious model did not differ from the less

Table 8.1 Estimated results for both formal and informal banks

	Formal (a)	*(b)*	*Informal*
C	0.029	−0.097573	–
Size	0.049*	0.054*	0.089***
liquidity	0.823	0.751	1.036
Capitalization	−3.844***	−4.062***	−8.421***
MPR	−0.025**	−0.017	−0.021
Growth		−0.002	
Inflation		0.014	
Gross capital formation		−0.004	
R^2	0.69	0.68	

NB: ***, (**), and [*] indicate significant at 1%, (5%), and [10%] respectively
Source Authors computation from estimation (2021)

parsimonious model in terms of values for the adjusted R^2, , signs and variables that showed statistical significance. In particular, model (a) and (b) reaches the best goodness of fit with an adjusted R^2 of about 69 and 68% respectively and according to the Akaike and Schwarz criteria model (a) is the preferred one, hence we proceed with the discussion using model (a).

The results showed that variations in loan supply of the universal banks are significantly influenced by the size of the bank, the capitalisation, and policy rate. A positive significant coefficient of the bank size suggests that an increase in the size of a bank leads to increases in credit. The estimates suggest that a unit change in the size of a bank causes a 0.05 percentage change in the loans of banks. This evidence collaborates with theoretical predictions and studies such as Kashyap and Stein (1995, 2000) which showed that small banks may find it more difficult to raise external funds in times of monetary tightening, hence they are expected to suffer disproportionately. Consequently, they may be forced to reduce their lending relatively more than large banks, which results in the size of bank to positively impact loan variation.

Similarly, Kashyap and Stein (2000) and Ashcraft (2001) showed that bank liquidity insulates banks from the effect of monetary tightening. For example, more liquid banks can draw down on their liquid assets to buffer their loan portfolios, hence they are unlikely to reduce lending in response to monetary tightening. Also, smaller banks' holding of liquid assets could be sensitive, thus, their lending could be sensitive to monetary policy. Further, lending of highly leveraged banks is expected to be more responsive to monetary policy than lending of well-capitalised banks (Kishan & Opiela, 2000). This evidence is found in this study where the capital structure and liquidity negatively and positively respectively affect loan variations, though liquidity is found to be statistically insignificant. The estimates suggest that a unit change in the capital structure, and liquidity position causes 3.84 and 0.82 percentage change in the loans of banks respectively.

Also, the effect of policy rate on loan variation is found to align with theory. An increase in policy rate is expected to inversely affect credit change, thus a positive change in policy rate directly affect retail rates which consequently affect the credit demand inversely. The estimate shows that a unit change in policy rate causes 0.02 percentage change in loans of banks, and the relation is negative. We also estimated the impulse response function of loan variation, capital structure, and bank size to one standard deviation innovation in monetary policy rate. The response function of loan variation, capital, and size to innovation in policy rate is shown in Fig. 8.2. We observed from the figure that apart from loans that respond in a decreasing manner, liquidity, size, and capitalisation respond by increasing. Also, we observed that on average, the full impact of policy change is between two to four years.

The above results for the formal banks stand in contrast to the evidence for the informal banks. The result in Table 8.1 column 3 indicates that monetary policy is insignificant in determining variations in credits of informal banks. This is expected because informal credit is not associated directly with central bank policy, since the informal bank is not under the required reserve regulatory framework—the route via which central bank policy influence the banking sector.

The above findings are not unexpected given the different regulatory framework that faces the two types of banks. As indicated in Sect. 2.3, the universal banks are regulated under the Bank of Ghana Act 2002 (Act 612) Amended 2016, which among others requires the universal banks to keep reserves with the Bank of Ghana. Through the required reserve, the BoG can perform its function as a lender of last resort, hence able to use its monetary policy rate to influence the retail rates of the universal banks. Contrary, the informal banks are regulated under non-bank financial Institution Act 2008 (Act 774), which does not require them to keep

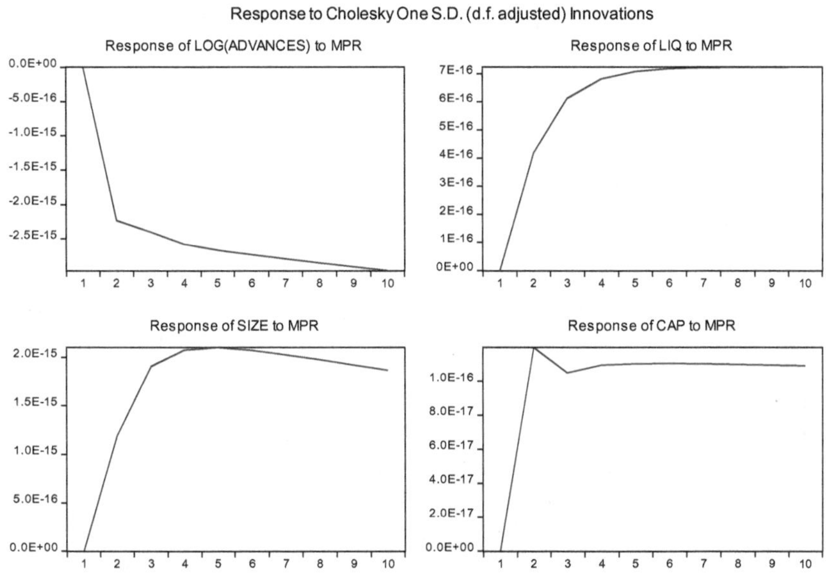

Fig. 8.2 Impulse response function to MPR shock (*Source* Authors' compilation from estimation)

reserves with the Bank of Ghana. This notwithstanding, both types of institutions compete in the credit market rather than complementing.

In summary, the study established that monetary policy affects the credit variations of the formal and informal banks differently. While monetary policy rate affects credit variation of formal banks significantly at a percentage point of about 0.025, it is insignificant in influencing the credit variations of the informal banks. The findings may be attributed to the different regulatory frameworks facing them, though they both operate in a common credit market (Fig. 8.2).

8.5 Conclusion and Recommendation

In conclusion, the study found contrasting evidence on the role of monetary policy in determining credit variations of formal and informal banks. Though the evidence in terms of directional impact appears to be in unison, informal banks credits are not significantly responsive to policy rate variations but formal banks do. Despite the lack of significance of policy variations on informal banking, the possibility of a dampening effect is likely given the increasing size and growth of the informal financial market. In the era of COVID-19 where central banks are stimulating growth by offering favourable policy rates, the informal banking activity could be leveraged or enhanced since the nature of work and economic activity is becoming more formal yet intrinsically informal.

The implication of the findings raises concerns about the right policy mix and effectiveness of monetary policy control in Ghana. The study has shown that relying solely on the monetary policy rate as an anchor to implementing monetary policy will only affect the universal without influencing that of the informal. Given the significant role of informal banking in the credit intermediation, a tool that abstract from them may cause monetary policy not to be able to achieve its' desired results. Consequently, we recommend the central bank adopt plethora of policy tools including unconventional monetary tools that may have direct influence on the informal market.

References

Adeusi, A. A. (2012). Effect of financial liberalisation on the performance of informal capital market. *Research Journal of Finance and Accounting*, *3*(6), 63–77.

Altunbaş, Y., Fazylov, O., & Molyneux, P. (2002). Evidence on the bank lending channel in Europe. *Journal of Banking and Finance, 26*(11), 2093–2110. https://doi.org/10.1016/S0378-4266(02)00201-7

Angeloni, I., Kashyap, A. K., Mojon, B., & Terlizzese, D. (2003). *Monetary transmission in the Euro area: Does the interest rate channel explain it all?*

Aryeetey, E. (1995). *Filling the niche: Informal finance in Africa*. East African Educational Publishers.

Aryeetey, E. (2005). Informal finance for private sector development in Sub-Saharan Africa. *Journal of Microfinance, 7*(1), 13–38.

Aryeetey, E. (2008). *From informal finance to formal finance in Sub-Saharan Africa: Lessons from linkage efforts*. In Paper presented at the High-Level Seminar on Africa Finance for the 21st Century Organized by the IMF Institute and the Joint Africa Institute. Tunis, Tunisia.

Aryeetey, E., & Gockel, F. (1991). *Mobilising domestic resources for capital formation in Ghana: The role of informal financial sectors*. African Economic Research Consortium, Paper 3.

Ashcraft, A. B. (2001, July 23). Do tougher bank capital requirements matter? New evidence from the eighties. *New Evidence from the Eighties*.

Ashcraft, A. B. (2006). New evidence on the lending channel. *Journal of Money, Credit, and Banking, 38*(3), 751–775. https://doi.org/10.1353/mcb.2006.0037

Asiamah, T. A. (2017). *Trends and determinants of household use of financial services in Ghana* (Doctoral dissertation). University of Ghana.

Aysan, A. F., Disli, M., & Ozturk, H. (2018). Bank lending channel in a dual banking system: Why are Islamic banks so responsive? *The World Economy, 41*(3), 674–698.

Bernanke, B. S., & Blinder, A. S. (1988). Credit, money, and aggregate demand. *The American Economic Review, 78*(2), 435–439.

Bernanke, B. S., & Gertler, M. (1995). Inside the black box: The credit channel of monetary policy transmission. *Journal of Economic Perspectives, 9*(4), 27–48.

Besley, T., & Levenson, A. R. (1996). The role of informal finance in household capital accumulation: Evidence from Taiwan. *The Economic Journal, 106*, 35–59.

Black, L. K., Hancock, D., Passmore, W., & Black, L. (2010). *The bank lending channel of monetary policy and its effect on mortgage lending.*

Carpenter, S. B. (1999). Informal credit markets and the transmission of monterary policy: Evidence from South Korea. *Review of Development Economics*, 323–335.

Debondt, G. J. (2013). Credit channels in Europe: A cross-country investigation. *PSL Quarterly Review, 52*(210). https://doi.org/10.13133/2037-3643/10926

Eberhardt, M., & Teal, F. (2010). *Productivity analysis in global manufacturing production.*

Ehrmann, M., Gambacorta, L., Martínez-Pagés, J., Sevestre, P., & Worms, A. (2003). Financial systems and the role of banks in monetary policy transmission in the euro area. In *Monetary policy transmission in the Euro Area: A study by the eurosystem monetary transmission network* (pp. 235–269).Cambridge University Press. https://doi.org/10.1017/CBO9780511492372.018

Ehrmann, M., Gambacorta, L., Martínez-Pagés, J., Sevestre, P., & Worms, A. (2001). Financial systems and the role of banks in monetary policy transmission in the euro area. *Available at SSRN 356660.*

Favero, C., Giavazzi, F., & Flabbi, L. (1999). *The transmission mechanism of monetary policy in Europe: Evidence from banks' balance sheets.* https://doi.org/10.3386/w7231

Foley-Fisher, N., Ramcharan, R., & Yu, E. (2016). The impact of unconventional monetary policy on firm financing constraints: Evidence from the maturity extension program. *Journal of Financial Economics, 122*(2), 409–429.

Gambacorta, L. (2005). Inside the bank lending channel. *European Economic Review, 49*(7), 1737–1759.

Gertler, M., & Gilchrist, S. (1993). The role of credit market imperfections in the monetary transmission mechanism: Arguments and evidence. *The Scandinavian Journal of Economics*, 43–64.

Ghosh, S., & Kumar, R. (2015). Monetary Policy and Informal Finance. *Economic and Political Weekly*, 18–20.

Guirkinger, C. (2008). Understanding the coexistence of formal and informal credit markets in Piura. *Peru. World Development, 36*(8), 1436–1452.

Hoff, K., & Stiglitz, J. (1990). Imperfect information and rural credit markets: Puzzles and policy perspectives. *The World Bank Economic Review, 4*(3), 235–250.

Huang, Z. (2003). Evidence of a bank lending channel in the UK. *Journal of Banking & Finance, 27*(3), 491–510.

Hung, F. S. (2015). Tax evasion, financial dualism and economic growth. *Academia Economic Papers, 43*(2), 175.

Igoni, S., Onwumere, J. U. J., & Amaewhule, P. W. (2020). Digital currency and monetary policy in the South African economy: Praxis of financial dualism and suggestion for Nigerians. *Asian Journal of Economics, Finance and Management*, *2*(1), 49–57.

Kashyap, A. K., & Stein, J. C. (1997). The role of banks in monetary policy: A survey with implications for the European monetary union. *Economic Perspectives-Federal Reserve Bank of Chicago, 21*, 2–18.

Kashyap, A. K., & Stein, J. C. (1995, June). The impact of monetary policy on bank balance sheets. In *Carnegie-Rochester conference series on public policy* (Vol. 42, pp. 151–195). https://doi.org/10.1016/0167-2231(95)00032-U

Kashyap, A. K., & Stein, J. C. (2000). What do a million observations on banks say about the transmission of monetary policy? *American Economic Review, 90*(3), 407–428. https://doi.org/10.1257/aer.90.3.407

Kashyap, A. K., & Stein, J. C. (2007). *7. Monetary Policy and Bank Lending* (pp. 221–262). University of Chicago Press.

Kashyap, A. K., Stein, J. C., & Wilcox, D. W. (1992). *Monetary policy and credit conditions: Evidence from the composition of external financ*e.

Kishan, R. P., & Opiela, T. P. (2000). Bank size, bank capital, and the bank lending channel. *Journal of Money, Credit and Banking, 32*(1), 121. https://doi.org/10.2307/2601095

Levenson, A. R., & Besley, T. (1996). The anatomy of an informal financial market: Rosca participation in Taiwan. *Journal of Development Economics, 51*, 45–68.

Liang, M.-Y. (1998). A note on financial dualism and interest rate policies: A loanable funds approach. *International Economic Review, 29*(3), 539–549.

Madestam, A. (2014). Informal finance: A theory of moneylenders. *Journal of Development Economics, 107*, 157–174.

Matousek, R., & Sarantis, N. (2009). The bank lending channel and monetary transmission in Central and Eastern European countries. *Journal of Comparative Economics, 37*(2), 321–334. https://doi.org/10.1016/j.jce.2008.09.008

Morduch, J. (1995). Income smoothing and consumption smoothing. *The Journal of Economic Perspectives, 9*(3), 103–114.

Myint, H. (1985). Organizational dualism and economic development. *Asian Development Review, 3*(1), 24–42.

Okurut, F.N., Margaret, B., & Ashie, M. (2004). Microfinance and poverty reduction in Uganda: Achievements and challenges. *Economic Policy Research Centre.*

Oliner, S. D., & Rudebusch, G. D. (1996). Monetary policy and credit conditions: Evidence from the composition of external finance: Comment. *The American Economic Review, 86*(1), 300–309.

Oliner, S. D., & Rudebusch, G. D. (1995). Is there a bank lending channel for monetary policy? *Federal Reserve Bank of San Francisco Economic Review, 2*(3), 20.

Peek, J., & Rosengren, E. (1995). Bank regulation and the credit crunch. *Journal of Banking & Finance, 19*(3–4), 679–692.

Pesaran, M. H. (2006). Estimation and inference in large heterogeneous panels with a multifactor error structure. *Econometrica, 74*(4), 967–1012.

Pruteanu-Podpiera, A. M. (2007). The role of banks in the Czech monetary policy transmission mechanism. *Economics of Transition, 15*(2), 393–428.

Sanfilippo-Azofra, S., Torre-Olmo, B., Cantero-Saiz, M., & López-Gutiérrez, C. (2018). Financial development and the bank lending channel in developing countries. *Journal of Macroeconomics, 55*, 215–234. https://doi.org/10.1016/j.jmacro.2017.10.009

Sarpong-Kumankoma, E., Osei, K. A., & Ladime, J. (2013). *Determinants of bank lending behaviour in Ghana essays on central bank independence in Africa View project determinants of bank lending behaviour in Ghana* (Vol. 4, Issue 17). www.iiste.org

Shem, A. O., & Atieno, R. (2001). Financial dualism and financial sector development in low income countries. *Savings and Development, 25*(4), 413–438.

Sichei, M. M. (2005). *Bank-lending channel in South Africa: Bank-level dynamic panel date analysis*. http://www.up.ac.za/up/web/en/academic/economics/index.html

Stein, J. C. (2012). Monetary policy as financial stability regulation. *The Quarterly Journal of Economics, 127*(1), 57–95.

Stiglitz, J. (1989). Markets, market failures and development. *The Journal of American Review, 79*(2), 197–203.

Walsh, C. (2003). Speed limit policies: The output gap and optimal monetary policy. *American Economic Review, 93*(1), 265–278.

CHAPTER 9

Revisiting MSMEs Financing Through Banking Reform Processes: Assessing the Ghanaian Legal Experiences

Francis Kofi Korankye-Sakyi, Omo Oyeniyi Abe, and Elijah Tukwariba Yin

9.1 Introduction

9.1.1 Background

A pathway to achieving growth in any economy is through access to finance, either for businesses or for consumers. In many Sub-Saharan African countries, the lack of adequate legal and institutional framework impedes securing credit financing for Micro, Small and Medium Enterprises (MSMEs). According to the International Monetary Fund (IMF) (2016), improved financial sector efficiency is considered critical

F. K. Korankye-Sakyi (✉) · E. T. Yin
Faculty of Law, College of Humanities and Legal Studies, University of Cape Coast, Cape Coast, Ghana
e-mail: francis.korankye-sakyi@ucc.edu.gh

E. T. Yin
e-mail: elijah.yin@ucc.edu.gh

J. A. Peprah et al. (eds.), *Financial Sector Development in Ghana*, Palgrave Macmillan Studies in Banking and Financial Institutions,
https://doi.org/10.1007/978-3-031-09345-6_9

for sustainable growth in Sub-Saharan Africa. It is noted to assist in mobilising deposits and source funding into productive sectors; for example, by providing start-up funds for prospective MSMEs (IMF, 2016). However, many developing economies are yet to realise the importance of, and contribution of MSMEs to the growth and development of an economy (Korankye-Sakyi, 2019). Literature is replete with banking reforms and implications for finance in particular and economies in general, however, there is a dearth of scholarly dimension to MSMEs which presents a lacuna that this chapter attempts to fill. It is observed that, by way of policy, there is little to account for as deliberate mechanisms to safeguard MSMEs at such critical times during banking reforms in the past. We highlight the substantial gap between the banking reform policies in Ghana from 1953 to 2016, and how they have reached the plights of MSMEs in accessing sustainable financing. It is against this backdrop that this chapter examines the institutional and regulatory framework underpinning Ghana's banking reforms and its impacts on access to credit by MSMEs. It seeks to give an up-to-date account of various facets of banking reforms and how they have evolved over the past decades under Ghana's financial development. The study used a qualitative desktop approach to assess how Ghana's banking reforms have been managed, with a comparative analysis of Nigeria and South Africa's successful reforms of their banking industries.[1] The analysis of the literature and data highlights how Ghana is performing compared with its neighbours in the sub-region in the areas of past reforms and impacts on financing MSMEs.

O. O. Abe
Faculty of Law, Huddersfield Business School, University of Huddersfield, Huddersfield, UK
e-mail: O.O.Abe@hud.ac.uk

[1] The IMF has indicated that with the exceptions of Ghana, Nigeria and South Africa with commendable financial sector developments worth appreciating, many sub-Saharan African countries have their financial markets only beginning to bud (IMF, 2016). This informed the choice of these countries for our comparative analysis.

9.1.2 *Banks and Banking Reforms*

Banks are institutions that serve as intermediaries to advance idle surpluses as a credit to businesses for economic growth. The banking sector, according to Akpansung and Gidgbi (2014) is traditionally regarded as an intermediary architecture through which transactions of deposits and payments for productive investments are routed. Likewise, Ojukwu-Ogba (2009) states that "the banking system in any given society is the artery through which the economic lifeblood of the nation runs". The consensus is that there is a direct correlation between banking sector reform and economic growth. It is therefore for no reason that the economic, social and human resource development agenda of any country is tied to the financial structure of that economy and its efficiency (Antwi-Asare & Addison, 2000).

According to Akpansung and Gidigbi (2014, p. 94), banking reform has been defined "as regular or irregular interception in rules and regulations guiding the operation of the financial institution, towards attainment of the international best standard, and sufficient backing of economic growth and development in a country". This confirms the theory and practices of those institutional reforms as an embedded administrative imperative to any progressive commercial jurisprudence (Yu et al., 2014). It also buttresses the assumption that commercial practices are associated with a high degree of risks as a result of the unpredictable trends in the world economic order, human factor errors and mismanagement (Lomnicka et al., 2005). Lomnicka et al., (2005, p. 26) argue that some of the pitfalls which confront banks in their paths are "unwise investments in questionable industrial projects, lending to countries with unstable economies, hazardous dealings in foreign currencies, and the investment of money received on short deposits in long-term transactions". These scholars, however, conclude that failures in the banking sector globally are not novel.

This backdrop, fundamentally, informs States to consistently engage in the reform of their banking laws in line with global best practices and as a means of fostering economic growth. Banking reforms become necessary, that is if need to keep in touch with economic reality. When a bank is managed unwisely or it naturally falls into unforeseen circumstances, crises arise (Lomnicka et al., 2005). Such unforeseen circumstances occurred during the financial crisis of 2008 (Ivashina & Scharfstein, 2010). The failure of the financial institutions in 2008 resulted in a freeze of global credit markets that required government interventions worldwide[2] (Ahamed & Mallick, 2019; Anginer et al., 2018; Berger et al., 2016; Ellul & Yerramilli, 2013). The effects of crashing of stock markets, capital squeeze, shrinking investment, decrease in income, loss of jobs, loss of homes, starvation and hunger culminating in economic meltdown were equally anticipated (Essien, 2011). Global South economies including Ghana were profoundly impacted and this was reflected in the financial and general economic growth in subsequent years.

Kayanula and Quartey (2000) have noted that "the dynamic role of small and medium enterprises (SMEs) in developing countries as engines through which the growth objectives of developing countries can be achieved has long been recognised". MSMEs are the livewire of the Ghanaian economy constituting approximately 85% of all registered businesses (International Trade Centre, 2016, p. ix). They contribute 70% and 49% to GDP and employment respectively (Muriithi, 2017; International Trade Centre, 2016, p. ix). Again, 85% of the MSMEs in Ghana offer employment in the manufacturing sector alone (Akorsu et al., 2012). This is a significant contribution from the sector that requires some deliberate protection and oiling to promote its further growth through effective government policy and legal frameworks. As would be discussed later in this chapter, the informal mobilisation of excess money in a form of cooperative organisations amongst the ordinary market vendors, artisans, farmers, casual labourers and individuals was given an impetus under the Financial Institution (Non-Banking) Law 1993 (PNDCL 328).

The remaining sections of the chapter are as follows: Section two discusses unsecured transactional financing and dangers to MSMEs in Ghana. Section three looks at the evolution of the legal regimes for banking reforms and application. Section four concentrates on the banking sector reforms in Nigeria and South Africa to underscore the comparative effects of banking sector reforms in other economies in Sub-Saharan Africa. Section five focuses on the current banking reforms and

[2] For example, the US government spent about $700 billion to bail out or insure the assets of several financial institutions. This was made possible through the *Troubled Asset Relief Program* (TARP) program of the United States government, designed to purchase assets and equity from financial institutions to strengthen its financial sector. It was signed into law by U.S. President George W. Bush on 3 October 2008. Likewise, the UK government declared a bank rescue package of about $740 billion in loans and guarantees.

implications for access to credit by MSMEs. Section six concludes the chapter with a rehash of the objectives, findings and recommendations.

9.2 Unsecured Transactional Financing and Dangers to MSMEs in Ghana

'Small Money Collection and Credit Agencies' (locally known as '*Susu*'[3] operators) served as a useful avenue for mobilisation of capital and for advancing small credit facilities to households and family businesses. As financial reforms broadened and the frontiers of banking and technology opened up the spheres of savings and credit, this practice has largely been extinct. This is also due to the spread of banking and savings houses reaching almost every corner of the country through the liberalisation and formalisation of the sector. Today, the core functions of these '*susu*' entrepreneurs and cooperatives are performed under the purview of the microfinance, savings and loans companies regulated under the Banks and Specialised Deposit-Taking Institution Act, 2016 (Act 930). These non-banking and deposit-taking institutions have played critical roles in the socio-economic stories of Ghana as their activities are mainly centred in rural and urban poor areas where the commercial banks have refused to establish there. Entrepreneurs in these institutions are themselves by definition MSME operators as indicated by their capital, number of employees and profit margins. To this end, ordinary citizens have had the faith to deal with them to start up or grow their businesses. These microfinance institutions (MFIs) have served as the retail financial sector of the economy. To many Ghanaians who constitute the 60% unbanked population (World Bank, 2018), their first-time experience in any formal banking setting is through these institutions.

In recent times, many cases of fraud have emerged in the industry. This has called for the whip of the Regulator to implement stringent regulatory and supervisory measures to curtail further illegalities to save depositors and customers from losing out their capital and economic wherewithal. In 2015, a major fraudulent cartel visited mayhem on their unsuspecting customers thereby causing a loss of billions of Ghana cedis

[3] In the local Akan dialect of Ghana "Susu" literally translates "to measure". The term was coined to describe a cooperative undertaking in which one receives 'a measure' of as much as what he contributes at the end of the month or year.

to customers who kept their monies with these companies. In these fraudulent schemes, DKM Diamond microfinance alone was estimated to have misappropriated about GH₵115 million of depositors' money (Oxford Business Group, 2019). The obvious action from the BoG was to close it down with 70 other microfinance companies which were involved in the same or similar financial offences. Again, the BoG was largely faulted for this national catastrophe for reneging on its governance functions; licensing, monitoring, supervision and sanctioning for non-compliance. To streamline the microfinance operations, the Central Bank increased the minimum paid-up capital for MFIs from GH₵500,000 to GH₵2 million whiles the Savings and Loans companies also saw a rise from GH₵300,000 to GH₵2 million. It was therefore not surprising that in the current reforms underway, over 400 such institutions have had their licences revoked as they failed to meet the minimum capital requirement deadline of 31 December 2018.

The overarching effects of this are the public's repulsion developed towards such institutions; and the subsequent negative incursions into the financial inclusion strides of government in the short and medium-term. The impacts of these revocations can only be felt much more by MSMEs than any other section of the economy. The apprehension is that the current situation may force small business owners to go back into the informal 'susu' era; which was considered unsafe, highly irregular and too risky. This will thwart the effort of the government at formalising the financial sector. Hopefully, if the 'susu' enterprise is reconsidered as part of future reforms and regulated properly, it may offer convenient avenues for MSMEs to have access to financial services. 'Susu' has more positive to offer especially to the large number of people who cannot afford or go through the official banking channels.

9.3 The Evolution of Legal Regimes for Banking Reforms and Application in Ghana

9.3.1 Phase One: 1953–1983

Since the Bank of Ghana (as it came to be known after independence) was established as the Central Bank of the nation in 1953, it has supervised phases of transitions and reforms of the banking sector till today under different institutional and legal frameworks. Between 1920 and 1950, the only banks that were in the Gold Coast (Ghana) were the Bank of British

West Africa and Barclays Banks. The Bank of Gold Coast was established by the colonial administration as a bank of issue, personal banking and business. The first reform was to split the bank into the Bank of Ghana (BoG) and the Ghana Commercial Bank (GCB). This afforded the BoG a focus as an independent central bank to operate as a bank of issue of currency whiles the GCB as the first indigenous commercial bank was to manage the public corporate accounts and curtail the dominance of the two foreign banks in the sector (International Institute for the Advanced Study of Cultures, Institutions and Enterprises, 2014).

By 1958, the Bank of Ghana supervised the issuing and management of the Cedi as the national currency of the country. In the same vein, the GCB was redirected to take up additional functions as government bankers; managing the accounts of all state agencies and departments. The BoG opened its first branches in the regions of Ashanti and Northern. The Agricultural Development Bank, the Merchant Bank and the Ghana Investment Bank were established between 1957 and 1965 as state-owned banks[4] to take care of agricultural growth, trade and mercantile banking and investment promotion respectively. The government of the First Republic had an overarching control in the financial sector with wholly or majority shares in the sector. This first phase of "reforms" was successfully managed and positioned the BoG to focus on its core mandate but little is recorded about the impact of the regime on the activities of MSMEs for policy decisions on financing.

Ghana's economic philosophy, however, started shifting towards a liberal market economy led by the private sector after the overthrow of Dr. Kwame Nkrumah's government in 1966. This truncated the monopolisation of trade, investment, service sectors, as well as the financial sector by the ubiquitous state.

9.3.2 Phase Two: 1983–2000 Reforms

Under the Provisional National Defence Council (P.N.D.C.) regime, the economic troubles at the time, beginning in December 1983, prompted the regime to succumb to the Economic Recovery Programme (ERP) of the International Monetary Fund (IMF) for economic assistance. As

[4] The political ideology of the government of the Convention Peoples' Party under the leadership of Ghana's first President, which leaned to the left was critical to such policy direction of the banking sector at the time.

part of the package that impacted the financial sector considerably, two key legislations were enacted; (1) the Investment Code of 1985(PNDCL 116) and (2) the Banking Law of 1989 (PNDC Law 225). The Investment Code supported the liberal school of thought that advanced the position to relax the restrictions on private investors' involvement in finance, trade and investment in the economy. To advance this course and provide the conduit for such investments, the Banking Law of 1989 was enacted to provide the milieu for effective and efficient banking, non-banking, securities and financial services. The PNDC Law 225 facilitated the involvement of local financial entrepreneurs to acquire operating licences to own and operate private banks in the country. In furtherance of the financial sector liberalisation, the government introduced both the Financial Sector Adjustment Programme (FINSAP) and the Financial Sector Strategic Plan (FINSSIP) which hiked savings and deposit trends culminating in deepening financial outlook and competition amongst the banking services providers (International Institute for the Advanced Study of Cultures, Institutions and Enterprises, 2014).

A major breakthrough in this phase was the passage of the Financial Institution (Non-Banking) Law 1993(PNDCL 328). The law was enacted for finance houses, building societies, discount companies, leasing and hire-purchase companies to be licensed to operate as non-banking financial operators. The advent of *'Susu'* which mobilised sums of money from market women, artisans and individuals daily as savings; and offered credit packages gained impetus under this legal regime. This provided an informal banking avenue and enhanced the financial inclusion profile of the masses for short- and medium-term financing of MSMEs.

In 1991, the government established the First Finance Company (a partnership between the Bank of Ghana and the Social Security and National Insurance Trust) with the mandate to help recapitalise ailing but viable companies in the financial sector. This was a specific state intervention to afford reliable and easier reach to credit by enterprises that suffered distress as a result of the ERP. As anticipated, these reforms helped to grow the banking sector of Ghana in the early 1990s with the surge in the rates of deposits by the citizens. The financial sector was largely deregulated. To a large extent, this phase brought along significant improvement in the banking industry, especially with the entry of the private sector into the business.

9.3.3 Phase Three: 2004–2015

In 2004, a new phase of the banking sector reforms was ushered in with the enactment of new banking legislation, Banking Act, 2004 (Act 673) (as amended by Banking Act, 2007 (Act 738). Act 673 (as amended) was passed to amend and consolidate the laws relating to banking, regulate institutions that carry on banking business and provide for other related matters (Banking Act, 2004 (Act 673) of Ghana). The provisions in the Act were to strengthen the licensing procedure, regulations, processes and sanctions regime in the banking industry (Part II, Act 673 2004). For example, to police the banks from deviating from their core businesses and avoid venturing into subsidiaries that will encourage capital diversion, the Act disallowed any bank from undertaking any commercial, agricultural or industrial venture unless it did that with such subsidiary of the bank explicitly registered for such other business. Before this law was the enactment of the Bank of Ghana Act (2002) Act 612 established as the central bank of Ghana (Section 3(2), Act 612) to help stabilise the general level of prices and support the general economic policy of government through effective and efficient operation of banking and credit systems in the country (Sect. 3(1), Act 612).

The essence of these legal regimes was to continue the reform process to streamline the industry and strengthen the banks to offer proper and secured financial services to the public and customers. Act 673 eliminated the phenomenon of secondary reserves and increased the minimum capital requirements of the banks with a capital adequacy ratio of 10%.[5] From an initial GH₵60 million in 2007, the threshold increased to GH₵100 million by 2013. Universal banking licences which allowed a bank to offer various banking products were issued for the first time to banks in Ghana under this law. Due to increased minimum capital requirements and the profitability of providing universal banking services, some banks went into mergers and acquisitions to stay in business and maximise profits. The various regulatory and legislative steps taken during this phase are considered key to making it a period that saw the banking sector interfacing with the general public more than any period of Ghana's history.

[5] The capital adequacy ratio (CAR) is a measure of a bank's solvency.

9.3.4 The Present Phase: 2016-Date

The 2015 Bank of Ghana report indicated that at the end of December 2015 before the present reforms began, there were 29 commercial banks comprising 12 indigenous Ghanaian banks and 17 expatriate banks. These banks operated in 1173 branches across the country with 62 non-banking financial institutions (NBFIs), 139 rural and community banks as well as 546 microfinance institutions. To bring the sector up to speed and redirect its mandate to avoid a financial meltdown in the economy, the current regime of reforms began with the enactment of the Banks and Specialised Deposit-Taking Institution Act, 2016 (Act 930). In line with the general mandate of the BoG, the recent normalisation exercise of the bank to sanitise the industry is deemed well-grounded and comes within the dictates of the law. The BoG is seen as revisiting its regulatory and supervisory framework to re-align the banking industry to be solvent with adequate capitalisation to achieve a strong and resilient base for the industry (Moody's Analytics, 2019). Even though the nature and manner of the programme have come under scrutiny, it has been embraced generally as timely and respected by the players and stakeholders in the sector.

With the coming into force of Act 930, every financial or banking practitioner was to anticipate a new paradigm to regularise the sector and prune it from excesses that were making the economic anchor fragile and precipitous to financial crises. Banking crises come along with collateral damages with locked-up investments; with dire rippling effects on sensitive economic actors and vulnerable agencies.

In 2018, the BoG undertook some regulatory and supervisory measures within the scope of its mandate, pursuant to Section 123 of Act 930 of 2016. By Directive (BG/GOV/SEC/2017/19) issued on 11 September 2017, a new minimum capital requirement for the commercial and universal banks at GH₵400 million from GH₵120 million was announced to be met by the close of 31 December 2018. Upon the failure to comply with this policy among other reasons,[6] five banks (Unibank, the Royal Bank, Sovereign Bank, Beige Bank and

[6] Apart from liquidity challenges and the inability of the banks to meet their CAR, other factors the BoG cites for the clean-up include poor corporate governance, false financial reporting and insider dealings amongst top executives and board room members of the banks. The UT and Capital banks are always cited as examples of banks that fell due to corporate governance malpractices and insider dealings as part of their shortfalls.

Construction Bank) were closed down on insolvency grounds and consolidated to form a national bank known as the Consolidated Bank of Ghana Ltd. About GH₵450 million was allocated out of an issued bond of GH₵5.76 billion by the government to capitalise on the CBG (Price Waterhouse Coopers, 2018). A year before the merger of the five banks, the UT and Capital Banks were liquidated, placed under receivership and eventually taken over by the GCB bank, which also has the government as the majority shareholder. It must be suggested that the government's hold on commercial banks should not be the desired preference of a market economy and the government must be keen to dilute its shares in these banks for effective private sector participation when the process has been stabilised in the shortest possible time.

In October 2018, the BoG released the names of 30 banks plus the ARB Apex bank,[7] with 1,546 branches in April 2018 as licensed banks in good standing to operate in the country. In its first-quarter report for 2019, which spells out developments and performance of the banking sector, BoG revealed that the number of licensed banks in good standing had reduced to 23 with 1.225 branches nationwide as of April 2019 (Bank of Ghana, 2019). This reduction is, again, due to the fact that the banking institutions have been affected by the ongoing reforms by way of receivership and liquidation, consolidation or closure; regrettably comprising mostly of indigenous banks, microfinance institutions and savings and loans entities. Again, in line with Section 11 of the Non-Bank Financial Institution Act, 2008 (Act 774), rural and community banks which fall under the ARB Apex Bank were given the final notice to recapitalise to meet the required minimum capital of GH1 million by end of February 2020 to stay in business. To this day, this directive to the rural banks has not seen its fullest implementation. Three (3) new banks have come up as a result of various mergers and acquisitions; Omni Bank and Bank Sahel Sahara have merged, First National Bank and GHL Bank have merged and the First Atlantic Merchant Bank Limited and Energy Commercial Bank have merged, after meeting the new minimum capital requirement. To resuscitate distressed indigenous banks, the government of Ghana has given sovereign guarantees through the Parliament of Ghana to the Ghana Amalgamated Trust Limited (GAT), a special purpose holding company financed through private pension funds in Ghana to inject fresh

[7] The ARB Apex bank is the umbrella bank established to manage all rural and community banks in Ghana.

equity capital in five indigenous banks. Consequently, the ADB Bank, the NIB Bank, the merged Omni/Bank Sahel Sahara, the Universal Merchant Bank and the Prudential Bank have benefitted from this vehicle after proven solvent with satisfactory corporate governance practices.

Alagidede et al. (2013) posit that the structure of any financial system determines how its monetary administration is carried through. Experts have alluded that in most emerging economies, where there are weak financial systems, MSMEs are confronted with significant hurdles in accessing capital from financial institutions (Chardon, 2014). It is found that the lapses in good corporate governance, the challenges of insolvency and insider trading among the failed banks leading to the revocation of their licences were real. By far, the current ongoing reforms becomes the most comprehensive and far-reaching in the history of the banking sector in Ghana. To support this position is to emphasise that the reform exercise was fundamental. From the evidence, it appears the BoG seems to be undertaking this process in good faith, having the good of the economy and its drivers (the businesses) in mind.

It must be noted that the implications of such reforms are hydra-headed and must be managed effectively with some reasonable level of caution. Recent developments, as a result of the reforms and 'normalisation' exercise in the sector, have given room for concerns leading to the expressions of fears about; the sustainability of the sector, the competent management of the rippling effects of the exercise, the erosion of the citizens' confidence in banking and savings, the further ramifications on MSMEs to access deposits and credit, and the ability of the Central Bank to hold on to its monetary role without succumbing to political dictates in order not to dislocate the economy. Clearly, effectively managed banking reforms under the suitable legal arrangement(s) tend to bring more positive outcomes than negatives. Many have argued that Ghana has a plethora of laws to govern its banking sector, but there is a lack of institutional structure to effectively implement these laws. There is the absence of clear, adequate and efficient legal provisions in the laws to regulate the banking sector. For a long time, this has been the quagmire often exploited by the players leading to the abuse and misapplication of the legal framework. A single, comprehensive banking law that will effectively and sustainably address the identified problems is a panacea to the uncertainties under the current Ghanaian banking laws.

9.4 Banking Sector Reforms in Nigeria and South Africa

The Nigerian banking sector has seen tremendous reform strategies, even though recent volatility affecting its entire financial sector tends to affect the industry in the long run. The structural reforms of the banking industry in Nigeria began almost in the same period as in Ghana. (Iganiga, 2010). The reforms in Nigeria's banking sector formed part of the total financial sector package undertaken under the 1986 Structural Adjustment Programme (SAP) (Akpansung & Gidgbi, 2014). The deregulation of the banking sector under the SAP in the 1990s served as a motivation for a relaxed regime that saw a record high licensing of more banks in Nigeria causing banking misfortunes in Nigeria (Ojukwu-Ogba, 2009).

Before the passage of the Banking Ordinance Act of 1952 that created the Central Bank of Nigeria (CBN) in 1958, unethical and banking irregularities malpractices characterised the sector affecting the performance of the sector for a long time. The Central Bank Act of 1958 gave the CBN supervisory and regulatory control over banks and other financial institutions.

Subsequently, the autonomous status of the CBN was strengthened with a wider mandate by the passage of the Banks and Other Financial Institutions Act 1991 (As amended in 1997, 1998 and 1999). In July 2004, the CBN announced a '13-point reform agenda' to regulate and streamline the banking sector as part of the process to reform the industry (Ojukwu-Ogba, 2009). The Banks and Other Financial Institutions Act, (Cap B3 Laws of the Federation of Nigeria, 2004) and the Central Bank of Nigeria Act, (Cap C4 Laws of the Federation of Nigeria, 2004) constitute the two principal laws regulating banking business in Nigeria among other laws were passed. Additionally, the CBN Act, of 2007 has come to repeal the CBN Act of 1958 and Banking Act of 1969 as further reform measures. Before this period, state-owned banks were dominant in the sector.

As part of the robust reforms in the Nigerian banking sector, the Price Waterhouse Coopers (2018) reports that the key measures implemented included the following: First, the CBN in 2005 specifically caused the banks to recapitalise which resulted in the rise of the minimum capital requirement from ₦1 billion to ₦25 billion for existing banks whiles the new banks were required of ₦2 billion. As a result, there has been a significant increase in the capital base of the local banks, a deepening in financial undertakings and propelling positive financial development indicators, which are inspiring confidence in the public leading to a 3.4% growth in savings (Akpansung & Gidgbi, 2014). The banks in Nigeria over the period have the capital base to finance huge capital-intensive projects and investments. This is attributed to the sector's increase in total assets from ₦3,753.3 billion in December 2004 to ₦4,515.1 billion in December 2005 (Price Waterhouse Coopers, 2018). Second, the CBN took legitimate steps to reduce the number of banks to 25 from a total of 89. Studies have found that this measure of a significant reduction in the numbers has accounted for the improvement in the credit extension to productive sectors of the Nigerian economy with a significant portion going to the MSMEs (Akpansung & Gidgbi, 2014). The credit support emanating from that windfall to the MSMEs is worth noting.

Following these measures, there is no doubt that the Nigerian banking sector was transformed and its banks compete on the continent as some of the stronger banks provide viable and efficient services beyond their territory. The so-called 25 megabanks now extend their operations beyond Nigeria into countries such as South Africa, Ghana, the UK, the USA, Benin and others (Ojukwu-Ogba, 2009).

Following the Nigerian experience, the likely future can be envisioned after the process is completed in Ghana. The likelihood of having an otherwise stronger banking sector running into a serious financial downturn can never be ruled out as long as strong institutional and regulatory frameworks are not promulgated to forecast, address and mitigate threats in time. It can be deduced that Ghana's recent reforms are akin to the Nigerian exercise and the logical expectation is that institutional limps would not be weakened by unforeseen hands to drop the process down along the way.

South Africa's banking sector is considered the most robust in Sub-Saharan Africa with significant market capitalisation. Apart from key sectors such as agriculture, mining, manufacturing and other financial services, the banking sector plays a key role in contributing to the growth of its GDP. The South African Reserve Bank (SARB) has oversight responsibility for the regulation and prudential management of the banks and allows the Financial Sector Conduct Authority (formerly, Financial Services Board [FSB]) to regulate the market conduct of financial service providers.

South African banks remained resilient in the face of the global economic crisis of 2008. Despite this fact, the SARB continued to comply with the Basel Capital Accord III to sustain the requirements for capital adequacy on the banks, provide an enduring regulatory regime, provide risk management policies, equip the sector against liquidity challenges and curtail financial malfeasance.

After 2008, various banking regulatory reforms started aimed at financial stability, consumer protection, managing credit exposure, financial inclusion and prudent banking operations. First, as part of the regulatory reforms, the National Credit Act (2005), Act 35 targeted at reforming the credit system was enacted. Again, the Financial Advisory and Intermediary Services Act (2002), Act 37 was also enacted to regulate the matters and operations of financial institutions.

The South African banking sector operates on the "Twin Peaks" model of banking and financial market regulation which discourages a sectoral approach to managing financial institutions in favour of legislative harmonisation to coordinate the overlapping activities of financial institutions (Global Legal Insight, 2021, p. 2). This system has been given legal backing with the coming into force of the Financial Sector Regulation, 2017 (Act 9) which also gives additional power to the SARB by the establishment of the Financial Stability Oversight Committee. Unlike Ghana, there are no fully fledged state-owned commercial banks in South Africa. The Postbank and the Land Bank for example are not registered to operate fully as operational banks. The Banks Act currently allows the Postbank and the Land Bank to operate under some exemptions which do not give them banking licences to operate as fully fledged Banks (van Zyl & Lopes, 2018).

The natural consequences of the deregulated sector have propelled takeovers and mergers resulting in large banks with varied interests in portfolios and financial services. The Standard Bank, Nedbank, First National Bank and ABSA with over 95% of domestic banking assets constitute the four major banks in the country. It is the regulatory structure in South Africa's banking system that will inform the basis for our policy recommendation to move Ghana away from the currently fragmented structure in banking governance. Due to the structure of the banking system of South Africa, it has over the years supported its MSMEs with a range of financial services.

9.5 Current Banking Reforms and Implications for Access to Credit by MSMEs

The obvious consequences of the banking reforms in Ghana over the years and their impacts on credit flow to the productive sectors of the economy cannot be overemphasised. The surge in competition in the sector is a result of the influx of foreign banks as well as the increase in the indigenous banks. The competition has mainly been keen on the size of deposits as well as the volume of shares of these banks on the market. The ultimate beneficiaries of the expansion are the MSMEs which have the option to select competitive bankers and negotiate for affordable credit conditionalities. The International Institute for the Advanced Study of Cultures, Institutions and Enterprises (IIASCIE) (2014) shares that competition is a catalyst to efficiency. This has manifested among the banks with significant improvement in service delivery to the banking population. Benefits of technological innovation in the areas of e-banking, telephone banking, SMS banking and the operation of automated teller machines (ATMs) have seen remarkable efficiency.

Liquidity in the financial sector has relatively been guaranteed with the support from offshore funds of expatriate banks to cushion and sustain the credit demand on the sector efficiency (IIASCIE, 2014). In effect, capital for investments has been relatively fluid while dependence on domestic capital mobilisation has been relaxed to cater for infrastructure development. With the availability of loans with competitive terms, MSMEs have the urge to take credit and structure their portfolios to ease their businesses. The majority of the banks have adopted business-friendly approaches to woe MSMEs into savings and access loans due to the competition in the industry.[8] This gesture has not been fully appreciated by the MSMEs because no significant impacts have been made concerning the increase in the volumes of credit needed as the lifeblood of these businesses. Again, mergers and acquisitions in the industry occasioned by the requirement for increased minimum capital have led to the creation of larger banking institutions with substantial capital bases and

[8] The majority of the banks in Ghana have dedicated desks with specialised officers to provide banking services to MSMEs, except that not so much can be said of better incentive packages to the MSMEs concerning the terms and conditions associated with the credit packages to these enterprises.

better balance sheets that can be relied upon to provide funding for major investments and support for economic growth.

Banking reforms, as noted throughout the various key reform phases in the country, are associated with an increase in the minimum capital of the institutions which is critical to the extent that banks are risk-averse and need better insulation during unexpected periods of liquidity challenges (Price Waterhouse Coopers, 2019; Bank for International Settlements, 2010). Generally, larger banks can mobilise and absorb the shocks within this period while many smaller and usually indigenous banks liquidate or close down. The swift responses by the government from 2016 to 2018 to the vulnerabilities in the financial sector are therefore necessary and may have to continue to sanitise the sector.

In contrast to the above positives, the negative impacts of the banking reforms on smaller banks and non-banking institutions present major challenges to MSMEs. The primary business of banks is to take deposits from the public and invest them in ventures to accrue profits to reimburse the customers with interest on demand according to the terms of deposits (Joachimson v Swiss Bank, 1921). The dissolution of a bank, therefore, comes with a great cost to such depositors and customers of the bank (Lomnicka et al., 2005). The eventual victims of such collapse are the individual account holders and the business community which is largely made up of MSMEs in Ghana. The fall of banks prompts financial panic in the system which instigates a run on the banks with the pressure no bank can mitigate. Panic situations have been identified as the effect of waning trust and confidence in the banking industry borne out of the information and experiences encountered by the banking public (Joachimson v Swiss Bank, 1921).

These ramifications have led to job losses as a result of the inability of MSMEs to expand and grow arising out of the financial challenges. We anticipate the likelihood of more businesses folding up due to the inflexibility of credit conditionalities that come with the few surviving banking institutions that may offer little credit to be accessed among competing industries including larger multinational companies (MNCs).

Ghana's bankable adult population stood at only 40% in 2014, while less than 20% had a formal savings account (World Bank, 2018). This is expected to reduce woefully due to the recent happenings in the sector which has forced citizens out of the banks by default of the process or by voluntary closure and withdrawal of accounts. The fear of losing capital or the sheer panic created by the incessant rumours of liquidation of these financial houses partly account for these panic withdrawals and closure of accounts.

Finally, we noted that the critical issue of credit to MSMEs which is the major theme for this chapter has been the worst affected under the current reforms. According to the BoG's own credit conditions survey in May 2020, the overall credit disbursement to enterprises (comprising mainly of MSMEs) has declined against projection. This development was attributed *inter alia* to the resultant closure of some banks and projected credit repayment challenges (BoG, 2020, p. 13). Instructively, data from the BoG suggests that the decline in credit to businesses has adversely impacted economic activities over the period (BoG, 2020, p. 13). Given this, it is expected that the BoG will in the future put adequate measures in place to streamline its policies to avert further credit tightening to MSMEs due to the impacts of its banking reforms.

9.6 Conclusion

The objective of this chapter is to discuss the institutional and regulatory framework underpinning Ghana's banking reforms and its impacts on access to credit by MSMEs. It also examines the various banking reforms introduced by the BoG since 1953. It aims at a comparative analysis of Nigeria and South Africa's successful banking reforms with Ghana to draw useful policy lessons.

9.6.1 Findings

This chapter finds that, first, the economic engine of every country is its MSMEs through which jobs are created, wealth distributed to the very commoner of the society, and to eradicate financial inequality. MSMEs also afford an economy the conduit to grow its GDP sustainably and fortified. Notwithstanding, MSMEs suffered the dire effects of the banking reforms that have ever taken place in the country; predominantly with liquidity constraints to finance their operations.

The deregulation of the banking sector in Ghana has not yielded the desired outcome to benefit the growing but struggling MSMEs population as was also in the case of Nigeria until strict liquidity measures were instituted. Banks willing to enter into the sector must be certified to have the requisite liquidity and the technical expertise to be licensed.

It is also noted that unlike in Nigeria and South Africa, the banking reforms in Ghana over the years and their impacts on credit flow to the productive sectors of the economy have not been significant. The surge in competition in terms of the size of deposits and shares of these banks on the market has lent little to the financing of MSMEs.

Additionally, the ineffective legal framework is noted as constituting a major avenue for banking sector mismanagement or irregularities due to the sophisticated and diverse players in the industry.

9.6.2 Recommendations

First, a good policy direction must anticipate a constant revision of regulations and updating of existing legislation to address emerging changes. Considering the changing nature of the economy, a constant revision and harmonisation of regulations and legislation in the banking sector are recommended.

Second, the banking sector policy of the country must focus on effective mechanisms and infrastructure for proper monitoring, supervision and compliance to keep growth balanced to play a critical role in supporting the MSMEs. This must be anchored on strong institutions, promulgating sound legal and regulatory frameworks, and widening financial inclusion mechanisms to reach more MSMEs. It is commendable that the Corporate Insolvency and Reconstruction Act, 2020 (Act 1015) has been promulgated which has repealed the Bodies Corporate (Official Liquidations) Act, 1963 (Act 180). Act 1015 is to provide a proper legal milieu to manage the administration and official winding-up of insolvent companies, including banking institutions and for related matters. It is one legislation resulting from the pitfalls identified with the recent banking reforms which have rather compromised many MSMEs negatively as against forecast and anticipation. However, the realisation of the new law's objects will depend largely on the key issues of monitoring, supervision and compliance.

Third, the imposition of strict regulations to check the entry of new institutions into the sector must be a key institutional policy approach. There is the need to evolve and apply measures that monitor banks to meet their liabilities to ensure the stability of the banks. A sound banking institution is theorised to be constituted the one that has the required liquidity and has readily funds to pay for current demands (Ellinger et al.,

2005). By sound inference, a bank without adequate margins cannot be allowed to operate in a sound banking environment.

Fourth, legal certainty is considered an essential tool for fostering a conducive and efficient financial milieu for sustainable socioeconomic development via formidable and sound MSMEs. To achieve this as part of a long-term financial sector policy development, it is expected that a single regulatory entity is established as the 'Financial Services Authority of Ghana' and empowered by transferring all regulatory authority concerning banking, investment business and insurance from the BoG to it. This will end the regime of having different regulatory agencies with separate legal regimes overseeing separate financial services entities in Ghana. This must be backed by a legal arrangement under a financial service and market regulatory law. It is suggested that a consolidated legal framework of all banking and related activities' laws is promulgated to be under the authority of a separate statutory body to oversee its compliance and implementation. This "Financial Services Act"[9] must look at prudential rules, financial stability, consumer finance and rights under retail banking, market efficiency, competition and compensation rules, international banking standards, compliance and enforcement among other far-sighted provisions for a stabilised financial market.

References

Literature

Ahamed, M. M., & Mallick, S. K. (2019). Is financial inclusion good for bank stability? International evidence. *Journal of Economic Behavior & Organization, 157*, 403–427.

Akorsu, P. K., Agyapong, D., & Kwashie, C. (2012). An alternative model for financing SMEs in Ghana. *International Journal of Arts and Commerce, 1*(5), 136–148.

Akpansung, A. O., & Gidigbi, M. O. (2014). Recent banking reforms in Nigeria: Implications on sectoral credit allocation and economic growth. *International Journal of Business and Social Science*, *5*(13).

Anginer, D., Demirguc-Kunt, A., Huizinga, H., & Ma, K. (2018). Corporate governance of banks and financial stability. *Journal of Financial Economics, 130*(2), 327–346.

[9] This is the authors' construct of a suggested nomenclature of such legislation.

Antwi-Asare, T. O., & Addison, E. K. Y. (2000). *Financial sector reforms and bank performance in Ghana*. Overseas Development Institute.

Baah-Boateng, W., Nketiah-Amponsah, E., & Alagidede, P. (2013). The Ghanaian economy: An overview. *Ghanaian Journal of Economics, 1*(1), 4–34.

Bank of Ghana. (2019). *Banking sector report*. https://www.bog.gov.gh/banking_sect_report/banking-sector-report-november-2019/

Bank of Ghana. (2020). *Banking Sector Developments, 5*(3), 1–18.

Basel Committee. (2010). Basel III: A global regulatory framework for more resilient banks and banking systems. *Basel Committee on Banking Supervision*.

Berger, A. N., Imbierowicz, B., & Rauch, C. (2016). The roles of corporate governance in bank failures during the recent financial crisis. *Journal of Money, Credit and Banking, 48*(4), 729–770.

Chardon, D. (2014, December 22). *ICC Endorses UNCITRAL Convention on the Assignment of Receivables in International Trade*. https://iccwbo.org/media-wall/news-speeches/icc-endorses-uncitral-convention-on-the-assignment-of-receivables-in-international-trade/CIio

Ellul, A., & Yerramilli, V. (2013). Stronger risk controls, lower risk: Evidence from US bank holding companies. *The Journal of Finance, 68*(5), 1757–1803.

Essien, E. B. (2011). The current global economic crisis: Gleaning from history. *Current Research Journal of Economic Theory, 3*(2), 69–75.

Global Legal Insight. (2021). Banking regulation 2021: South Africa. https://laworld.com/law-news/global-legal-insights-banking-regulation-2021-the-south-african-chapter/

Iganiga, B. O. (2010). Evaluation of the Nigerian financial sector reforms using behavioural models. *Journal of Economics, 1*(2), 65–75.

International Monetary Fund. (2016). Financial development in Sub-Saharan Africa: Promoting inclusive and sustainable growth; Prepared by Montfort, M, Larry, C. et al., vii.

International Trade Centre. (2016). *Trade impact for good*. Retrieved from https://sustainabledevelopment.un.org/content/documents/10275International%20Trade%20Centre%20contribution.pdf

Ivashina, V., & Scharfstein, D. (2010). Bank lending during the financial crisis of 2008. *Journal of Financial Economics, 97*(3), 319–338.

Kayanula, D., & Quartey, P. (2000). The policy environment for promoting small and medium-sized enterprises in Ghana and Malawi. *Working Paper Series, Finance and Development Research Programme*, 15, University of Manchester.

Korankye-Sakyi, F. K. (2019). *Factoring as means of promoting small and medium scale enterprises: The case for a legal framework for credit factoring in Ghana*, LLM Thesis: University of Pretoria.

Lomnicka, E., Hooley, R., & Ellinger, P. (2005). *Ellinger's modern banking law* (4th ed.). Oxford University Press.

Moody's Analytics. (2019), BOG publishes update on status of banking sector reforms in Ghana. *Regulatory News*. https://www.moodysanalytics.com/regulatory-news/Jan-04-19-BOG-Publishes-Update-on-Status-of-Banking-Sector-Reforms-in-Ghana

Muriithi, S. (2017). African small and medium enterprises (SMEs) contributions, challenges and solutions. *European Journal of Research and Reflection in Management Sciences, 5*(1).

Ojukwu-Ogba, N. E. (2009). Banking sector reforms in Nigeria: Legal implications for the banker–customer relationship. *Commonwealth Law Bulletin, 35*(4), 675–686.

Oxford Business Group. (2019). Strong prospects remain for Ghana's banking sector despite slowing growth. *Ghana Financial Services Overview*.

PriceWaterHouseCoopers. (2018). Ghana banking survey. Having Secured the New Capital; What Next for Banks?

PriceWaterHouseCoopers. (2019). Banking reforms so far: Topmost issues on the minds of bank *CEOs*, Ghana Banking Survey.

The International Institute for the Advanced Study of Cultures, Institutions and Economic Enterprise. (2014). *The changing face of banking in Ghana and the implications for Ghana's economy*. Retrieved from https://www.interias.org.gh/position-paper/changing-face-banking-ghana-and-implications-ghana%E2%80%99seconomy

Van Zyl, C., & Lopes, R. (2018). Is South Africa Ready for a state-owned bank? *SA Financial Markets Journal*. https://financialmarketsjournal.co.za/is-south-africa-ready-for-a-state-owned-bank/

World Bank. (2018). *Financial inclusion report*. https://www.worldbank.org/en/news/press-release/2018/04/19/financial-inclusion-on-the-rise-but-gaps-remain-global-findex-database-shows

Yu, J. S., Hassan, M. K., Mamun, A., & Hassan, A. (2014). Financial sectors reform and economic growth in Morocco: An empirical analysis. *Journal of Emerging Market Finance, 13*(1), 69–102.

Statutes

Banking Law. (1989). PNDC Law 225.

Banking Act. (2004). Act 673 (Amended by Banking Act [2007]. Act 738).

Banks and Other Financial Institutions Act. (2004). Cap B3 Laws of the Federation of Nigeria.

Banks and Specialised Deposit-Taking Institution Act. (2016). Act 930.

Banking Ordinance Act. (1952). Nigeria.

Bodies Corporate (Official Liquidations) Act. (1963). Act 180.

Central Bank of Nigeria Act. (2004). Cap C4 Laws of the Federation of Nigeria.

Corporate Insolvency and Reconstruction Act. (2020). Act 1015.

Financial Institution (Non-Banking) Law. (1993). PNDCL 328.
Financial Advisory and Intermediary Services Act. (2002). Act 37 of South Africa.
Financial Sector Regulation. (2017). Act 9 of South Africa.
Investment Code. (1985). PNDCL 116.
National Credit Act. (2005). Act 35 of South Africa.
Non-Bank Financial Institution Act. (2008). Act 774.

Case Law

Joachimson v Swiss Bank. (1921) 3KB 110 (CA) at 127.

CHAPTER 10

Effectiveness of Credit Risks Management Policies Used by Ghanaian Commercial Banks in Agricultural Financing

Abraham Nyebar, Adefemi A. Obalade, and Paul-Francois Muzindutsi

10.1 Introduction and Background

10.1.1 Background to Agricultural Finance in Ghana

The significance of economic growth and development cannot be overemphasized. For example, agricultural contribution to GDP is between 45 to 50% while about 80% of Ghana's population are engaged in

A. Nyebar · A. A. Obalade · P.-F. Muzindutsi (✉)
School of Accounting, Economics and Finance, University of KwaZulu-Natal, Durban, South Africa
e-mail: MuzindutsiP@ukzn.ac.za

A. Nyebar
e-mail: anyebar@pentvars.edu.gh

A. A. Obalade
e-mail: ObaladeA@ukzn.ac.za

J. A. Peprah et al. (eds.), *Financial Sector Development in Ghana*, Palgrave Macmillan Studies in Banking and Financial Institutions,
https://doi.org/10.1007/978-3-031-09345-6_10

agriculture in 2020 (MoFA, 2021). Despite its strategic importance, agriculture requires sustainable financing methods. In this study, agriculture financing is explained as the total credit granted to the agriculture sector by commercial banks to stimulate agricultural activities for sustainable economic growth (Amadhila & Ikhide, 2016; BoG, 2021). Agriculture financing comes in various forms (BoG, 2021). Successive Governments in Ghana have tried to increase agricultural financing through agricultural programs such as the provision of farm inputs to farmers (MoFA, 2020). Agricultural financing can also be in a form of government guarantees on loans granted for agricultural purposes, imposition of quotas on loans extended to farmers, and subsidized interest rates on specific forms of loans granted to the agricultural sector (Amedi et al., 2019); institutional credit schemes received by Ghana from donor countries such as the USA and Canada in the form of agricultural credit and technology support for farmers to promote agricultural activities serve as forms of agricultural financing (Amedi et al., 2019); and credit facilities granted by Ghanaian commercial banks for agricultural activities (BoG, 2020).

However, agriculture financing comes with high credit risk (Misra, 2021) and lending to the sector has been challenging due to the high credit risk associated with it (BoG, 2020). An instance of credit risk encountered by commercial banks in agriculture financing includes; counterparty default risk which has been referred to as the possibility of defaulting in the repayment of principal and interest by borrowers on credits granted (Apanga et al., 2016). Consequently, the Ghana Poverty and Inequality Report (GPIR, 2016), posited that only 8% of rural Ghanaians can access credits to invest in agriculture. Similarly, the banking sector report of the Central Bank of Ghana (BoG, 2021) indicated that the agriculture sector has inadequate financing as compared to the other sectors of the economy due to the high default rate associated with the sector. Table 10.1 presents the trend of sector distribution of loans by the Bank of Ghana (BoG) from 2016 to 2020.

Apart from the mining and quarrying sector, the agricultural sector accounts for the lowest share of the credit from 2016 to 2020 ranging from 3.6% in 2016, 4.1% in 2017, 4.7% in 2018, 5.9% in 2019, and as low as 3.7% in 2020 (BoG, 2017, 2018, 2019, 2020, 2021). The service, commerce, and finance account for the highest share of the credit within the period. The information in Table 10.1 suggests that commercial banks have in a way neglected the agricultural sector as a result of the high credit risk associated with the sector.

Table 10.1 Distribution of credits by sector (%)

Sector	*2020*	*2019*	*2018*	*2017*	*2016*
Miscellaneous	7.0	12.19	10.1	10.8	8.9
Service	26.6	24.1	21.5	21.1	18.9
Transp. & communication	8.4	8.0	8.7	8.4	9.1
Commerce and finance	24.6	20.9	24.5	25.2	24.7
Elect, water & gas	7.1	6.3	8.6	8.3	15.2
Construction	9.6	8.2	7.8	11.0	9.9
Manufacturing	10.7	10.9	10.9	8.1	7.6
Mining and quarrying	2.5	2.9	3.2	3.0	2.0
Agric, forest and fishing	3.7	5.9	4.7	4.1	3.6

Source BoG (2017, 2018, 2019, 2020, 2021)

The neglect in the agricultural sector is because, commercial banks have over the years employed inadequate CRM practices to manage risk associated with agriculture financing though the sector has been exposed to high lending risk (BoG, 2021). Despite the low distribution of credit to agriculture, Table 10.2 shows that the proportion of non-performing loans (NPL) in the agricultural sector was significantly and on averagely higher between 2016 and 2020 as compared to the other sectors. NPL ratio in the agricultural sector was 42% in 2016, 43.6% in 2017, 35.5 in 2018, 23.1 in 2019, and 19.8 in 2020 (BoG, 2017, 2018, 2019, 2020, 2021) indicated in Table 10.2.

Table 10.2 Proportion of non-performing loans in each sector (%)

Sector	*2020*	*2019*	*2018*	*2017*	*2016*
Miscellaneous	6.5	3.7	7.6	6.8	5.2
Service	9.4	9.9	12.8	14.3	13.8
Transp. & communication	7.7	9.9	9.4	19.8	5.7
Commerce and finance	17.4	17.8	18.8	30.9	27.5
Elect, water & gas	25.8	28.3	37.8	31.3	13.6
Construction	22.6	17.9	18.6	21.3	9.7
Manufacturing	18.2	17.7	24.5	27.8	25.6
Mining and quarrying	11.7	12.2	7.2	18.0	20.9
Agric, forest and fishing	19.8	23.1	35.5	43.6	42.0

Source BoG (2017, 2018, 2019, 2020, 2021)

CRM is an emerging current phenomenon among commercial banks in agriculture financing even though commercial banks lack adequate or effective practices of managing them (Lagat et al., 2018). In addition, CRM is key to the survival of Ghanaian commercial banks in agriculture financing (Boateng, 2019). To this end, many commercial banks in Ghana have created CRM departments responsible for managing credit risk (Kessey, 2015). The CRM practices adopted by these departments involve the effective implementation of CRMPs and embracing the best strategies to mitigate credit risk associated with agriculture lending (ADB, 2020; GCB, 2020). On the whole, the inability to efficiently and effectively implement CRMPs and strategies to mitigate credit risk associated can result in liquidation or bankruptcy as recently witnessed in Ghana.

As a result of credit risk-related distress, UT and Capital Banks were liquidated in 2017. Beige, Sovereign, Unibank, Construction, and Royal banks were also liquidated while the Heritage Bank and Bank of Baroda also exited the market on their own decision (BoG, 2018). This prompted the attention of the Central Bank of Ghana the need to stabilize the banking sector. The collapse of the banks led to an increase in the capital requirements of banks to GHS400m ($83 m) indicated in the 2017 Ghana banking sector report. These banks were liquidated partly because of their inability to effectively implement CRMPs to manage credits given to the agriculture sector. Most of the loans taken by clients for their agriculture activities could not be recovered and put these banks into serious liquidity problems. GN bank was also later liquidated in 2019 (BoG, 2020) for similar reasons. One major issue appearing in the central bank report is the inability to manage credit risk (BoG, 2018). The gravity of credit risk associated with non-performing loans in the agriculture sector has caused commercial banks to increase their interest rates in lending to the sector.

To mitigate credit risk and its attendant effect on banks' profitability performance, effective practices are required. Although commercial banks have credit policies, some of them might not be effective or have implementation challenges. The main objective of this study is to evaluate the effectiveness of the implementation of CRMPs of Ghanaian commercial banks in agricultural financing. In an emerging economy such as Ghana where agriculture is the mainstay of the economy, specific attention must be directed at the formulation and implementation of effective CRMPs to encourage agricultural lending as a viable business. Few studies (see e.g., Asiama & Amoah, 2019; Owusu-Boafo et al., 2020; Wubin et al., 2020) in Africa have focused on agriculture lending even though commercial banks continue to battle with the associated credit risk. There is therefore a need to investigate the effectiveness of CRMPs of commercial banks in agricultural financing. To the best of the researchers' knowledge, the subject matter has not been adequately explored to reduce risk exposure. This chapter is designed to take a broad investigation into the effectiveness of CRMPs of commercial banks in agriculture financing. Unlike the few studies on CRM available in Africa, the combination of thematic and document is perceived as offering the correct inferences on CRM strategies. This study informs regulators of the need to closely monitor commercial banks on the implementation of risk management policies to minimize credit risk associated with agricultural lending.

10.1.2 Regulatory and Institutional Background of Credit Risk Management in Ghana

Though the CRM problem is becoming more significant in policy debates, research on CRM in developing economies such as Ghana is in its infancy. Precisely, evidence on CRM issues in Sub-Saharan Africa and Ghana, in particular, remains scant notwithstanding the conception that recent credit risk problems in the banking system pose a significant threat to the health of both developed and developing economies (BoG, 2020). The Bank of Ghana, in its efforts to ensure that Ghanaian banks comply with the more demanding provisions of the Basel II (1999), initiated the CRM promulgation of the Banking Act, 2004 (Act 673), which was passed by the Parliament in October 2004 to replace the Banking Law 1989 (PNDCL 225). Further, the Bank of Ghana encouraged all Ghanaian banks to adopt the full provisions of Basel II with slight modifications by December 31 2009 (BoG, 2010). This notwithstanding, credit risk has received little attention in the Ghanaian context. The Banks and Specialised Deposit-Taking Institutions Act, 2016, Act, 930, requires all Ghanaian commercial banks to fully operate as limited liability companies and provide agricultural credit supply while adopting measures to minimize credit losses. A judgmental pattern of Character, Ability, Means, Purpose, Amount, Repayment, Insurance, Interest, Commission, and Extras usually referred to as the CAMPARI, and ICE has been adopted by the banks to provide consistency in reporting and grouping attributes into categories to minimize credit risk. This suggests that the Ghanaian banking sector has a regulatory framework and credit risk reporting mechanism. However, it is not clear whether the current framework creates a conducive environment for the effectiveness of the implementation of credit risk management policies in agricultural finance. Additionally, existing credit policies might not be effective or may have implementation challenges when it comes to agricultural finance. Thus, a new study evaluating the effectiveness of the implementation of credit risk management policies in agricultural finance in the Ghanaian context will shed more light on this topic.

10.2 Literature Review

10.2.1 Conceptualization of Credit Risk

CRM can be contextualized by liquidity theory, which has a strong connection with credit risk resulting from borrowers' defaults which reduces cash flow and affects liquidity (Abdelaziz et al., 2020). Liquidity in this context is described as the ability and certainty to convert an asset back into cash within the shortest possible time (Pham, 2021). The theory implies that banks must hold a large number of liquid assets or readily marketable short-term liquid assets against possible demand and unforeseen circumstances (Pham, 2021). Lower liquidity negatively impacts agricultural lending (Maloba & Alhassan, 2019). The liquidity crisis of banks sparks interest in the role of the regulations that could promote resilience in banks (Mohd-Amin & Abdul-Rahman, 2020). Mohd-Amin and Abdul-Rahman (2020) argued that liquidity risk does not need to be covered by equity but an adequate volume of liquid assets. Thus, this chapter pays specific attention to liquidity risk in Ghanaian commercial banks.

In addition, the adverse selection theory was originated by Stglitz and Weiss (1981). The theory is underpinned by two main assumptions. Firstly, lenders cannot differentiate between the appetite of borrowers and risk levels. Secondly, credit contracts are subject to limitations (Stiglitz & Weiss, 1981). That is; where the debt responsibilities are more than the business or project proceeds, borrowers have no responsibility to pay out of their own pockets (Ahlin & Waters, 2016). Crawford et al. (2018) found that while increases in adverse selection increase loan prices and average defaults, reducing credit supply, the market power of the banks can mitigate these negative impacts. The adverse selection theory does not guarantee that the loan applicant secures credit when the interest is high, particularly at a time where loanable funds are insufficient (Vallee & Zeng, 2019).

Furthermore, credit referencing theory highlights information sharing which was earlier initiated by Freimer and Gordon (1965) and later extended by Stiglitz and Weiss (1981). The theory states that credit markets do not have adequate, complete, and reliable credit information on individuals (asymmetry of information). For a credit information system to be effective, there should be a reliable national identification system for individuals; an adequate regulatory or policy structure; a framework that supports information sharing, and a system that protects clients' privacy in Kenya (Saruni & Koori, 2020). The purpose of credit referencing theory is to expose any single borrower from a wide range of credit providers through credit bureau institutions. To establish a strong ground for the institution of credit referencing, the Ghanaian parliament has formulated the Credit Reporting Act, 2007 (Act 726) to regulate and guide the credit information of borrowers. However, Kusi et al. (2016) and BoG (2021) questioned the effectiveness of credit information sharing in Ghana and this necessitated the discussion of credit referencing in this chapter. These different theories explain the possible behavior of lenders and borrowers when faced with credit risk and provide assumptions to use in evaluating the implementation of CRM strategies.

10.2.2 Review of Empirical Studies on the Implementation of Credit Risk Management Strategies

The empirical studies covered the effectiveness of the implementation of CRMPs adopted by commercial banks to minimize credit risk in agriculture lending. Thus the empirical review provided a broader view on whether or not the implementation of CRMPs by commercial banks in agricultural lending is effective, and provided information on the implication of the implementation of CRMPs in commercial banks, following Afande (2014). The MoFA (2021) indicated that improper and poor bookkeeping by agriculturalists, lack of collateral, poor yield resulting in inadequate cashflows and the lack of insurance schemes for agricultural produces are some of the major challenges that hinder the full implementation of CRMPs. Owusu-Boafo et al. (2020) found that high credit risk resulting from poor implementation has negative consequences on banks' profitability. Focusing on loan appraisal in credit granting as one of the elements involved in the implementation of CRMPs, Adusei (2018) revealed that Ghanian commercial banks have departments responsible for the management of credit risk. In addition, Asiama and Amoah (2019) showed that monitory policy does not affect the growth of non-performing loans while poor credit appraisal negatively affects loan recovery. The findings of these studies (Adusei, 2018; Asiama & Amoah, 2019) suggest that banks could have CRM departments but might not have competent, experienced staff or well-resourced staff, who are regularly trained to properly appraise loans before they are granted.

Pham (2021) found that improper monitoring could lead to the diversion of loans for different purposes other than for agricultural use and can cause serious credit risk policy implementation challenges among banks. In the context of loan authorization and approval as CRMPs, Afande (2014), found that top management support in, loan authorization and approval greatly influenced CRM practices of commercial banks in Kenya. This is confirmed by Luvsannyam et al. (2021) in a study on non-performing loans recovery in Mongolia. Similarly, Afriyie et al. (2018) employed interview and document analysis and results proved that credit risk can be ameliorated through loan authorization and approval in Ghanaian commercial banks. It implies that agricultural loans that are not properly authorized and approved are likely to result in higher credit risk.

Empirical studies have also evaluated the role of 5Cs (Borrowers' Capacity, Character, Credit Worthiness, Credit History, and Credit Disbursement) in the implementation of the CRMPs of commercial banks in agricultural financing. The 5Cs, except for credit disbursement, are effective in minimizing credit losses and must be used by banks as methods for assessing borrowers (Caplinska & Tvaronaviciene, 2020). Also, BoG (2021) indicated that screening out for bad and good loan applicants and the capacity of agricultural borrowers to repay loans and interest were good signals for reducing credit risk associated with lending. Gouri and Mahajan (2017) examined different models of financing small farmers' agricultural value chain in India and found 5Cs as significant factors in mitigating credit risk. Evaluating the effect of compliance cost on the supply of bank credit to agriculture using a deferential approach, Moss and Suh (2020) concluded that 5Cs are effective in determining the extent to which agricultural borrowers comply with loan contracts. This finding was confirmed by Nguyen (2020) who found that collateral is inevitable in banks' lending. This implies that because commercial banks do not know the possible hidden information about the quality of borrowers, agricultural borrowers may change their attitude after receiving loans which can lead to credit losses. In this context, Flatnes and Carter (2019) in Tanzania and Odonkor (2018) in Ghana indicated that loan policies should be structured backed by collateral to ensure the prohibition of unsecured lending. This finding highlights the role of increasing competition among the banks concerning the strict implementation of the CRM process. This may be one of the dynamics characterizing the current commercial banking sector in Ghana and hence warrant further investigations.

In addition, some studies have also evaluated the use of the credit risk register and policy manuals in the granting of loans. The credit risk register is a document that contains detailed and comprehensive information on all the risks of commercial banks and should be regularly updated to measure the credit risk exposure of the banks (IMF, 2020). Risk register varies from bank to bank and serves as a source of information for reporting risks associated with commercial banks in agriculture lending (BoG, 2019, 2020). O'Har et al. (2017) studied the development of risk register tools in the United States and concluded that a risk register facilitates an organization's risk management governance and reduces risk. This was collaborated by Abbas et al. (2018) and Apanga et al. (2016) who indicated that banks should develop and use credit manuals to ensure consistency in credit decisions and ease lending processes to effectively mitigate credit risk in all types of lending. Monye et al. (2020) found that Nigerian banks' compliance with internal guidelines, laws, and regulations build effective credit reporting[1] systems and avoid credit risk exposure of banks. In the context of Ghana, BoG (2019) reported that the Central Bank of Ghana requires commercial banks to have an effective regulatory framework and environment to create the infrastructure of credit reporting to minimize credit risk exposure of the banks in agricultural lending. Therefore, the BoG supervises commercial banks to follow the internal guidelines laid down by the banks, comply with laws and regulatory requirements governing the financial sector to avoid credit risk exposure associated with agricultural financing (BoG, 2020).

Odonkor (2018) evaluated CRM in Ghana and concluded that banks in Ghana do not effectively enforce restrictive covenants and suggested

[1] Credit reporting has been described as the act of gathering information about the behavior and make such information available whenever a decision is to be made regarding a borrower's application for new credit (Monye et al., 2020).

that, borrowers should be made to sign bonds of good behavior when accessing credits for agriculture activities. Abu-Bakar et al. (2020) argued that corporate boards should ask for direct monitoring using covenants that allow them to intervene and exercise direct control of their interest to minimize credit risk. Considering risk transfer, the effect of possible losses from loans granted to borrowers in Ghana can be transferred to third parties through insurance as found by Odonkor (2018) in a study on Adansi rural bank in Ghana using descriptive statistics. Using a similar methodology, Dlugosch et al. (2018) evaluated the personality-based selection of entrepreneurial borrowers in reducing credit risk and revealed that loan losses would be minimized through insurance cover among commercial banks in Kenya. This finding is confirmed by Acharya et al. (2021) who carried out a study on credit lines and liquidity insurance channels in the United States and found insurance to be a key factor in minimizing credit losses. Based on this finding, the effects of credit risk would be easily transferred to third parties through insurance and prevents commercial banks from suffering entirely from credit losses associated with agricultural lending.

10.2.3 *Remarks on the Implementation of Credit Risk Management Policies in Ghana*

Commercial banks in Ghana have comprehensive CRMPs. Even though the CRMPs have been put in place by the commercial banks, these banks would continue to face CRM challenges if the policies are ineffectively implemented (BoG, 2021). However, commercial banks are faced with implementation challenges in agriculture financing resulting in loan losses and high credit risk (Ascui & Cojoianu, 2019). Therefore, there is a need for commercial banks to dramatically focus and improve upon their implementation practices to minimize credit losses and increase credit risk exposure reduction in agricultural lending reduce credit losses (Pham, 2021).

10.3 Methodology

10.3.1 *Population and Sample Selection*

The study adopted a qualitative approach involving the use of interviews and analysis of policy documents from BoG. There are 23 commercial banks in Ghana (BoG, 2021) representing the population of the study. The demographic profiles of the four commercial banks, namely GCB, ADB, SBG, and PBL, purposively selected for this study are given in Table 10.3. The four selected banks were most appropriate for this study because they are widely distributed over the country. They are the leading banks in terms of the provision of agriculture lending with a market share of 73% of the agriculture loan portfolio in Ghana (BoG, 2020; Ofori-Nyarko, 2017).

Ten (10) credit officers were purposively targeted in the selected four commercial banks for the interview. As a result of the nature of their jobs, the participants selected had relevant knowledge and could provide adequate information on CRM in agricultural lending. The Head Offices were purposively targeted for the interview because they oversee the implementations of risk management policies in all other branches and can best provide reliable and accurate data. Further, nine (9) farmers were purposively interviewed. The participants were purposively selected from an association of farmers with not less than five years of experience in farming. This is intended to gather information from the horses'

Table 10.3 Demographic profile of GCB, ADB, SBG, and PBL

Name	*Year established*	*Profile/Content*	*Branches*
GCB	1957	GCB was set up in 1957 to provide financial services and credit facilities to the indigenous agriculture business community in Ghana	185
ADB	1965	ADB bank was instituted in the 1960s by the Government of Ghana to increase access to agriculture financing. It is a universal bank offering a full range of banking products with more focus on agriculture	78
SBG	1999	Stanbic Bank began operations in Ghana in 1999 following the acquisition of majority shares in Union Mortgage Bank by the Standard Bank Group. The bank acquired a Universal Banking License in 2004. It also provides agricultural loans	40
PBL	1966	The Central Bank of Ghana also approved the establishment of more banks such as Prudential Bank (PPBL), an indigenous Ghanaian bank to strengthen and increase access to credits	43

Source BoG (2017, 2018, 2019, 2020, 2021)

mouths regarding their relationship with the banks in agricultural finance. The purposive sampling approach was not only cost-effective but was also time-effective.

The interview guide for the credit officers from the banks include; factors that are considered by Commercial Banks in agriculture lending; approval and authorization procedures in the granting agricultural loans; challenges faced by Commercial Banks in agriculture financing/lending; challenges faced by commercial banks in the implementation of CRMPs; roles played by experience and qualification in the appointment of credit officers; and possible best strategies that can be used by Commercial Banks to mitigate credit risk in agriculture financing. The interview for the farmers concentrated on challenges faced in agricultural lending. The basis of the interview guide was to uncover additional information on the implementation of CRMPs which could not be revealed through the document analysis. The interview lasted between 22 to 33 min for participants from the banks and 15 to 25 min for the farmers. The staff for the interview was considered based on the assumption that; the results supplement the information that was gathered through and policy documents summarized in Table 10.4.

The study used data on policy documents from the Central Bank of Ghana. Glenn (2009) described document analysis as a systematic procedure used to review, examine, and evaluate printed or electronic[2] documents. This procedure was followed and interpreted to provide meaning, understanding, and develop empirical knowledge. This provided background information on the implementation of CRMPs in commercial banks and contextualized the data gathered through in-depth interviews. The analysis involved skimming and thoroughly reading through the policy documents before the results were interpreted. The in-depth interviews were used to probe and respond to user data as it emerges. In line with Braun and Clarke (2006), thematic analysis was employed to enable the researcher to categorize and present the themes that were developed from the evidence collected; transcribe the interview results; read through the interview responses, and look for patterns or themes among the responses of the participants; familiarize with data gathered; assign preliminary codes to data to describe the content and search for

[2] Electronic documents represent both computer-based and internet-transmitted materials of an entity (Glenn, 2009).

Table 10.4 Details of documents analyzed

Policy	*Nature of policy document*	*Page*	*Year*
Credit agreement policy	Disclosure and product transparency rules for credit products and services	8, 13	2017
Credit guarantee policy	Disclosure and product transparency rules for credit products and services	9, 19	2017
Training and professional competence policy	Disclosure and product transparency rules for credit products and services	19	2017
Reporting requirements policy	Bank of Ghana notices to Banks, Non-Bank Financial Institutions, and the General public notice No. Bg/Gov/Sec/2011/04	2–6	2011
Examination and investigation policy	Risk management directives, BoG	7–14	2021
Knowing your customer policy	Credit reporting regulations, L. I 2394	1–20	2020
Minimum capital requirements policy	Requirements for banks and specialised deposit-taking institutions licenses under the banks and specialised deposit-taking institutions Act, 2016 (Act 930)	3–5	2016
Risk monitoring, and review policy	Risk management directives	7–14	2021

Source BoG (2017, 2018, 2019, 2020, 2021)

patterns of themes; review the themes, define main themes, and present key findings.

10.4 Results and Discussion

10.4.1 Analysis of Interview Results

As indicated in the methodology section, Results were obtained from the interviews and from analyzing policy documents on credit risk management from the Central Bank of Ghana. This section discusses the results generated from the interviews with credit offers and established farmers involved in borrowing to finance their agricultural activities. The discussion is divided into two sub-sections examining the responses from the credit offers and established farmers, respectively.

10.4.1.1 Interview Results from Credit Officers

An interview was purposely conducted for credit officers and managers. Participants from the commercial banks who availed for the interview were given coded names, namely A_1, A_2, and A_3, from Bank A; B_1, B_2, and B_3 from Bank B; C_1 and C_2 from Bank C; as well as D_1 and D_2 from Bank D. Some of the interview results reported on credit risk mitigation strategies are bank-specific. Regarding the factors considered by banks before granting agricultural credits, specific factors are mirrored by banks before approval and extension of credit to customers. Such factors could be an eye-opener to the bankers as to the potential risk inherent in the venture financed by the credit. From Bank A, the factors considered include: confirming whether the loan request is meant for agriculture purposes; confirming whether the borrower is into crop farming, livestock, agro-business, or any of the agriculture value chain activities; determining whether the business is legal and acceptable; examining the expertise of credit officers but does not consider the technical background of the officers; determining the reasonableness of the amount requested; and evaluating the repayment capacity of borrowers. A participant indicated the following vignette:

> We even take the qualification of the management. Do they have what it takes to run the kind of business they are talking about?

It was further stated that loans that are not supported by collaterals are risky and commercial banks are likely to suffer losses should borrowers default in repayment. A participant during a probe submitted the following:

> …there are rules around the securities. So… Bank of Ghana says that you have to give a security cover of 120%.

Other areas considered by the commercial banks before granting credits are the location, financial performance of the agricultural borrowers' businesses, and ready market. In Bank B, most of the factors identified in agriculture financing are similar to those identified by Bank A. A participant in bank B indicated that agriculture credit is appraised using basic pointers like the purpose of the loan and business cycles such as weather conditions if it involves crop farming. The participant further explained in the following verbatim words:

> ...so, assuming somebody comes to you, he wants to farm maize in the dry season. Is it possible? What has he put in place to assure that he can farm maize?

It was also found from Bank B that the competitiveness of a borrower; cost of operations and whether the borrower belongs to the association are significant factors. As a participant puts it in the following words:

> ...sometimes, you look at where the person is coming from. If he has an association, that is credible and the association can also guarantee for the person the bank will be okay to go ahead and finance the customer.

The interview results from Bank C and Bank D are not too different from the results obtained from Bank A and Bank B. The results from Bank C and Bank D showed that management qualities, collateral facility, the experience of credit officers, and cash flow of agricultural borrowers are some of the major factors considered in agriculture financing. Regarding loan authorization and approval process in commercial banks, the interview results revealed that authorization and approval processes in agriculture financing are almost the same in all the commercial banks as follows:

Borrowers approach the bank for agriculture credit; the business unit in the commercial banks put together a credit paper; relationship manager/office provides the customer with a checklist of the requirements; relationship managers interview the borrower to get more information on the cash flow, business plan, and products. The next stage is to pay a working visit to the agribusiness location to verify what is on the ground. Once the relationship officer is satisfied with the information on the ground, a credit paper is generated. From a participant, a credit paper is essentially a document that contains all credit information of borrowers. It includes the borrowers' credit history, cash flow, location, security, amount of loan requested, the purpose of the loan, interest charged, and the repayment period. The credit paper is forwarded to the risk department for full examination and recommendations. Apart from Bank B and Bank C that have agribusiness Units dealing with agricultural-related credits, Bank A and Bank D do not have such Units. It was further indicated that the risk department would then assess the loan request and make recommendations to the Credit Committee or the Managing Director or the Board based on the authorized limit for approval. The information is communicated back to the application station and then the borrower is informed of the outcomes. If the request is approved, the borrower is required to fulfill all pre-disbursement conditions before the loan is finally disbursed.

Regarding the challenges encountered in agriculture financing in a commercial bank, numerous of them were revealed. In Bank A, these include lack of dedicated units or officers that purposely handle agriculture financing; no specific or specialized unit set up to handle loans that are granted to the sector; inadequate attention given to the agriculture sector as far as financing is concerned and its associated high risk and poor yield. Poor yield in agricultural borrowers' businesses leads to poor performance of loans resulting in repayment problems. One of the participants indicated that there are instances where farms were granted loans and

> '...all of a sudden all the birds died' and 'where there was an issue with tilapia being poisoned'.

This implies that where poor yield is recorded, productivity reduces and affects the cash flow of agribusinesses thereby creating an inconvenient situation to repay loans with interest. Moreover, lack of proper documentation was cited as a big challenge to securing a loan. A participant state,

> ...assuming a farmer without proper documentation is applying for a loan in the dry season, you go and there's nothing on the field. How would you be able to tell whether this farmer is really into farming?

It implies that the poor record-keeping habit of the agriculturists is a major challenge in providing commercial banks with the necessary documentation required for loan approval. Explaining further, it was stated that,

> ...You know, the farmers usually don't keep books or good records on their operations

This implies that most agricultural borrowers do not keep proper accounting records and documentation. Results also showed that rainfall may fail farmers, thereby making the recovery of loans difficult. Even farmers might have a bumper harvest but might not have a market to sell their produce to repay their loans and these loans will grow bad affecting the performance of commercial banks. One of the participants indicated that,

> ...Farmers also don't have available markets. Just imagine someone taking a loan to go and farm or ...grow crops and tomatoes. These products are highly perishable. So the moment you finish and you don't get the ready market, you are not able to preserve them anyway. So, we should give you so much to invest in and how are we going to recover them?

In Bank B, the results of the interview are not too different from that of Bank A. One participant in Bank B added that crop production in Ghana is seasonal and borrowers specializing in crop farming might not have inflows for some particular seasons to repay their loans. Also, some of the agricultural borrowers divert the loans and begin to avoid credit officers as soon as the credits are disbursed. Further, participants pointed out the lack of long-term funding for the agriculture sector; inadequate agricultural infrastructure such as irrigation processes, storage facilities, and processing facilities post major challenges in loan repayment. Another major challenge also pointed out in Bank B is the fact farmers are not registered and do not have proper identity or information with the credit reference bureaus in Ghana. As one of the participants explained,

> ...it becomes difficult for their data to be captured in the credit reference bureau but if it is a commercial farmer that has proper identity you know that you can check that they have information on the date of birth, where they were born...

In Bank C, most of the interview results reported are similar to those reported by Bank A and Bank B. In Bank D, the interview results reported are similar to the same mentioned by participants in Bank A, Bank B, and bank C.

With regards to the experience and qualification requirements of credit officers, the interview results from participants of the four commercial banks generally revealed that commercial banks consider experience and qualification in the appointment of credit officers. A participant in Bank C stated that,

> … If I don't have the experience, I wouldn't know how to structure an agricultural paper. I wouldn't know palm (referring to a tree crop) takes this number of years to start uprooting. So let's say I give an oil farmer a loan and I asked him to pay me monthly, how can he pay?

It implies that, without experience, it is practically difficult for credit officers to restructure agricultural loans granted to borrowers. In Support, a participant in Bank D stated that,

> .. We train people on the job but when it has to do with a critical position fit for the purpose…, the criteria is not about somebody bringing you in from whatever political position…

The implication is that, if one does not have experience, he is not considered fit to work as a credit officer. It also implies that training on the job is necessary for developing the experience of loan officers who can manage credits advanced to agricultural borrowers and minimize losses. Adding, it was also found that,

> When we see or when the team see that you are more than capable of doing the job and you don't have the academic qualification we train you on the job

this means that people without any qualification who are capable of working as credit officers are appointed and trained on the job to manage credit-related transactions associated with agricultural financing.

10.4.1.2 Interview Results from Borrowers in Agricultural Finance

An interview was conducted for a section of 9 farmers (borrowers) purposively selected to get a better understanding of their testimonies about their capabilities and challenges relative to lending. For confidentiality purposes, participants were given coded names from F1 to F9. Results indicated that agriculturalists encounter serious challenges such as lack of access to agricultural loans; inadequacy of the loan; rainfall failures; lack of irrigation facilities; lack of ready market; lack of storage facilities during bumper harvests; inability to properly organize their books; lack of technical capacity; overreliance on traditional systems of farming (lack of innovation) and lack of successive governments' attention to subsidize the sector as a result of its associated high credit risk to stimulate growth. A participant indicated that,

> ...Governments over the years have supported agriculture on paper....No efforts have been made to put their ideas into practice...how can you play with the very sector that puts food on your table, provides your people jobs, and serves as the foundation of the economy

Results also indicated that commercial banks have not placed much significance on agricultural financing even though the farmers think they should be treated differently from the other sectors such as service and manufacturing among others. Lamenting, a participant stated;

> The banks have neglected us..they give us very little loans and expect high interest...so we are not able to expand to increase our income to pay the little loans given to us... we should be treated differently from other sectors like manufacturing and the service that do not border on life

Lack of storage facilities was also cited as a major challenge. Most of the farmers do not have storage facilities to store their produce from the farms. The perishable farm produces are sometimes left at the mercy of the buyers who frequently control the prices to offer to farmers. A participant recounts that;

> Most people in this community are tomato farmers..but we don't have houses to store our goods..so we take the buyers to the farms to buy,.. and at times at very cheap prices...if the buyers don't come, the products get rot and make us default in paying our loans.

The results further indicated lack of collateral, short repayment terms, and cash crop diseases as some of the challenges faced by farmers.

> ...yes cocoa and coffee have a lot of money...but I struggled to raise the initial capital because the banks failed me that I didn't have the collateral...Now, the banks give me loans because I can have some but it is limited to one year...I defaulted on the loan taken last year because of the disease which attacked the farm...

The implication is that banks cannot grant loans to farmers under longer repayment conditions. The amount will be scanty. Besides, shorter repayment conditions will put the farmers under high pressure to repay. Even though cash crop farming such as cocoa and coffee generates a lot of income, the proliferation of diseases can affect productivity and reduce the repayment power of borrowers.

10.4.2 *Review and Analysis of Policy Documents on Credit Risk Management*

The policies and guidelines concerning the supervision and monitoring of commercial banks for effective implementation of laid down policies, guidelines, and regulations were summarized in eight different themes discussed in subsequent sub-sections.

10.4.2.1 Credit Agreement Policy

The Credit Agreement Policy requires lenders to provide written copies of the credit agreement to borrowers. The written agreement produced by commercial banks must be signed by both the borrowers and authorized representatives of lenders (commercial banks). The credit agreement is expected to adhere to the required information, format, and any further amendments indicated in the Collateral Registry Rule established by the parliament of Ghana under the Borrowers and Lenders Act, 2008 (Act 773). The purpose of the Credit Agreement Policy is to create an enabling environment to promote easy access to credits and repayment of loans in commercial banks to reduce credit losses (BoG, 2020). The review indicated that the Credit Agreement Policy of the Bank of Ghana regarding the supervision of commercial banks was very detailed and effective. This policy implies that; all commercial banks were required to develop a Credit Agreement Policy with their borrowers before credits are administered. The policy also requires commercial banks to be very rational by requiring a secured property as collateral to minimize loan losses. The policy also suggests that credits would not be advanced if the commercial banks and borrowers do not agree, and reach a consensus.

10.4.2.2 Credit Guarantee Policy

The Borrowers and Lenders Act, 2008, Act 773, described credit guarantee as an agreement where a third party undertakes or promises to satisfy on-demand an obligation of a borrower in a credit facility or credit transaction to which this Act applies. By the Credit Guarantee Policy, borrowers are required to provide collateral security or guarantors as part of credit agreements before loans are granted as stipulated by the Borrowers, and Lenders Act, 2008, (Act 773). Guarantors, therefore, provide a guarantee for credits secured by borrowers to protect the banks against possible credit losses resulting from defaults. The Credit Guarantee Policy guidelines instituted by the BoG for commercial banks are very comprehensive and effective. Commercial banks that would comprehensively implement this policy are likely to reduce large credit losses as borrowers would mostly secure credits by credit guarantee provided by guarantors in good standing. The findings from the document analysis conformed to the interview results that agricultural borrowers were only granted credits by commercial banks upon the provision of collateral or a credit guarantee. This conforms to Wubin et al. (2020) who explored credit-granting schemes in Ghana using descriptive statistics and identified regular use of credit guarantee schemes as one of the efficient strategies of mitigating credit risk in agricultural financing. It is noteworthy that investors and farmers who engage in small and medium agribusiness lack the needed collateral to secure a loan for agriculture activities (BoG, 2021; MoFA, 2020). Also, it is most difficult for small and medium-term agribusinesses to attract guarantors as a result of their low-income nature and lack of a good asset base.

10.4.2.3 Training and Professional Competence Policy

Training and professional competence are very significant in minimizing credit risk in commercial banks. Training and competencies do not only increase cash flow but also most importantly expand the lifespan of the banks. Lenders such as commercial banks are required to train and develop the professional competence of their staff (BoG, 2020). To promote the development of staff training, and professional competencies to improve debt collection, lenders such as commercial banks must undertake continuous professional development activities to acquire relevant skills in managing credit-related risks (Kitonga, 2017). Consequently, the BoG requires commercial banks to assess the training needs of their staff to: ensure that staff members are appropriately trained; supervised, and also ensure that staff with a high level of competency are managed by the banks. Subsequently, Chowdhury and Alam (2020) in a study on the impact of ethics on the performance of commercial banks in Bangladesh postulated that commercial banks are required to establish a code of conduct and make it available to promote the professionalism and ethical standards of credit staff to minimize loan losses. The policy implies that commercial banks may encounter serious credit management problems if they do not have adequate and qualified credit officers to manage credits in agriculture financing. The policy further suggests that credit might easily be poorly managed if commercial banks do not have well-trained professionals to manage credits or if credit management is left in the hands of the wrong personnel. This is evidenced from the interview results which indicated that most of the credit officers of GCB, ADB, SBG, and PBL were mostly trained on the job. The interview results further showed that credit officers were recruited from various fields of study including psychology, accounting, sociology, economics, and political science among others to administer agricultural credits. This hinders the full implementation of this policy in minimizing credit losses in agricultural financing since credit might easily be poorly evaluated and managed. For instance, it will be difficult for a credit officer without agricultural training background to efficiently evaluate the prospects of farms for agricultural loans. The interview again revealed that some banks do not have technical units specifically responsible for agricultural credit administration and its associated high credit risk.

10.4.2.4 Reporting Requirement Policy

By section 93 of the Banks and Specialised Deposit-Taking Institutions Act, 2016 (Act 930), commercial banks are required to prepare periodic reports for supervision. The reporting requirement policy indicates that commercial banks are expected to report to the BoG weekly, monthly, quarterly, bi-annually, and annually. The reason for the reporting is to compare the short-term liquid assets of commercial banks with their respective short-term obligations to mitigate against future operating and credit losses (BoG, 2019). The monthly reports are required to cover critical areas including prudential reports, collateral returns, and institutional exposures. The prudential reports include summary information on deposits, current, and non-current assets, and liabilities. Essentially, the monthly reports to the BoG also cover reports on loans and advances giving detailed descriptions of both performing and non-performing loans by various commercial banks. Any anticipated effects resulting from credit losses in agriculture financing by the commercial banks would easily be determined through the monthly reports in minimizing credit risk exposure of these banks. The BoG requires commercial banks to report on collateral registration[3] for loans granted monthly. The reporting requirement policy initiated by the BoG for commercial banks is very detailed, reliable, and comprehensive but not adequate as BoG prudential requirements for commercial banks did not specifically include reports on agricultural lending and its associated high credit risk. These requirements did not give details of how loans given by commercial banks specifically for agricultural purposes would be reported and monitored. It is expected that proper books of accounts particularly relating to agricultural lending and its associated CRM are kept by commercial banks and should be well supervised and monitored by the Banking Supervision Department of BoG to minimize possible credit losses. This is missing in the policy document on reporting requirement of BoG. Regarding collateral, however, lenders such as commercial banks have no effective way of searching for charges[4] on properties that have been offered for credit. This is due to the several endless litigations, and a significant rate of non-performing loans which are usually recorded in the lenders' books of accounts.

10.4.2.5 Examination and Investigation Policy

It is expected that BoG examines all the commercial banks licensed to operate in Ghana. By section 94 of the banks and Specialised Deposits-Taking Institutions Act, 2016, (930), the BoG is required to examine the affairs and operations of each commercial bank operating in Ghana, at the frequency and time which it considers appropriate and prudent. The exercise covers the evaluation of the potential credit risk faced by each commercial bank in its operations. The requirements of section 94 of Act 930 on the examination of commercial banks' affairs and operations fall within the regulatory and supervisory controls of the BoG. The examination role of the BoG is statutorily required and not discretionary. The BoG may use the services of professionals such as lawyers and qualified accountants as agents to carry out the investigations on its behalf. This is very adequate and would likely result in reducing credit

[3] The Collateral Registry is a body established by Parliament under the Borrowers and Lenders Act, 2008 [Act 773] to principally register charges and collaterals created by borrowers to secure credit facilities provided by lenders (Borrowers and Lenders Act, 2008, Act 773).

[4] 'Charge refers to, mortgage, security, interest, lien, pledge, assignment by way of security, covenant, restriction, reservation, lease, trust, order, decree, judgment, title defect (including retention of title claim), or any other encumbrance of any nature other than liens arising by operation of law' (Borrowers and Lenders Act, 2008, Act 773).

losses in agriculture financing if commercial banks comply with all the policy requirements. However, this policy is general and does not specifically relate to agriculture lending despite the risky nature of agriculture financing and its inevitable credit risk which needs special attention. The policy favors the manufacturing, service construction, and commerce and finance industry where much attention is focused in terms of agricultural share of credits.

10.4.2.6 Knowing Your Customer Policy

The BoG has established the Knowing Your Customer Policy (BoG, 2021) as a requirement for commercial banks to comply to assess the credibility of borrowers as a means of minimizing credit losses. The committee on banking supervision (Basel, 2015) specifically published a paper on customer due diligence purposely for banks. The essence of the report was to enable banks to identify the relevance of ensuring the adequacy of controls, and procedures put in place to appropriately identify, and know very well the customers they deal with in everyday activities. The credit referencing bureaus play a significant role in providing credit information of borrowers for commercial banks to know their borrowers very well before advancing credits (BoG, 2020). The policy ensures that commercial banks in Ghana develop customer acceptance policies, and procedures that serve as guides and identify customers, and borrowers who are possibly likely to pose potential credit risk to the banks as far as lending is concerned (BoG, 2021). The policy discussed in this section would be very effective if it is well implemented by commercial banks to establish the credibility of potential borrowers in the agriculture sector through information sharing to reduce credit risk exposure (Brown et al., 2009). The findings from the analysis of this policy document conformed to the interview results, indicating that commercial banks must know all available information about their agricultural borrowers to avoid bad loans in agricultural lending.

10.4.2.7 Minimum Capital Requirement Policy

Banks are expected to meet the minimum capital requirement[5] (BoG, 2021). The minimum capital requirement of banks is to ensure the capital

[5] Capital requirement is the minimum amount of capital a bank or financial institution must hold as required by its regulator (BoG, 2019).

adequacy[6] of the banks (Madugu et al., 2020). The requirements are relevant to ensure that commercial banks do not take excess leverage and become insolvent in their lending to agricultural businesses stipulated by the Banks and Specialised Deposits-Institutions Act, 2016 (Act 930). The banking industry of Ghana has undergone three recapitalization programs in the last 17 years. In 2007, the stated capital of banks was increased from GH¢ 7 million to GH¢ 60 million, then in 2012, banks were directed to raise their stated capitals to GH¢ 120 million, and in 2017, the stated capital of banks was increased to GH¢ 400 million (Obuobi et al., 2019). The Bank of Ghana has established the minimum capital requirement policy of banks as part of its mandate to regulate, geared toward developing, strengthening, and modernizing the financial sector to advance the economic vision, and transformational agenda of the government to reduce potential credit risk (Obuobi et al., 2019). As a result, the bank of Ghana revoked the licenses of two insolvent banks (UT Bank and Capital bank) as a warning to the rest of the banking sector, and also to prepare the grounds for further reforms (BoG, 2020). This is similar to the views of Udom and Onyekachi (2018). Further, Rufo and John (2017) investigated the impact of credit risk on the capital adequacy of 567 banks in the Philippines. The study revealed that the capital adequacy ratio has negatively impacted the earnings of banks. Therefore, capital adequacy has been considered the best effective tool for ensuring banks' safety and soundness (Molefe & Muzindutsi, 2016). However, it is noted during the review that, there are no specialized banks with a focus on agriculture and with different capital requirements. All the commercial banks are given equal regulatory capital requirements without any special attention on the agriculture sector.

10.4.2.8 Risk Monitoring and Review Policy

The BoG is required to monitor and review potential risks faced by commercial banks, the source of the risk, the likelihood of occurrence, and the extent of the impact on the operations of the banks. The document analysis shows that the BoG has established the Board Risk Committee which provides advice to the Board on the overall presentation and expected future risk tolerance of banks. Importantly, this policy requires the Board Risk committee to regularly, and annually review,

[6] Capital adequacy is expressed as a ratio of equity as a percentage of risk weighted assets (Madugu et al., 2020).

and make recommendations regarding the risk management strategies undertaken by the regulated banks. This review shall include all policies, procedures, the models used as well as limits set to effectively reduce the credit risk exposure of commercial banks as stipulated by section 16 of the Banks, and Deposit-Taking Institutions Act 2016 (Act 930). However, findings from the analysis of this policy contradict the interview results which confirmed the lack of effective monitoring and review of credit transactions and credit risk associated with agricultural lending as borrowers were reported to be avoiding and hiding from credit officers after receiving agricultural credits. The interview further revealed that monitoring and review have become a challenge for commercial banks as a result of the scattered nature of the borrowers and its associated high monitoring cost. The findings contradict Nwude and Okeke (2018) who used the OLS regression model to evaluate the impact of credit risk on the performance of Nigerian banks and found that monitoring and review were very significant in minimizing credit risk. The findings also contradict Pham (2021) who indicated that effective monitoring is necessary for reducing credit losses and diversion of loans. This implies that the monitoring and review policy requirement of BoG for commercial banks is not well implemented. Additionally, there is no special risk monitoring and review policy specifically regarding agriculture financing in Ghana.

10.4.2.9 Comments on the Analysis of Policy Documents from BoG

The findings of the policy document analysis indicated that some of the policies were not very well implemented. The Credit Agreement Policy stressed the need for borrowers and lenders to reach a consensus through an agreement contract before credits are granted. The Credit Guarantee Policy seems to favor only agricultural borrowers with a large asset base to afford credits with collateral. This implies that the agricultural sector is of no strategic importance to the commercial banks contrary to BoG (2021) reports that indicated that, the agricultural sector is the foundation of the Ghanaian economy and needs special attention to grow. Further, the reporting requirement policy does not make provision for how transactions specifically relating to agricultural lending and its associated high credit risk should be reported. Similarly, the examination and investigation policy insists on regular examination, and supervision of commercial banks by the Bank of Ghana to minimize credit risk exposure. Besides, the 'know your customer' policy through information sharing has some lapses as Ghana is well noted for poor address systems, and that, the credit bureau institutions in Ghana are not very robust to capture all information about all borrowers. The minimum capital requirement policy does not have special consideration for agricultural lending considering the special nature of agriculture and its significance. Also, there is no detailed policy regarding CRM, particularly in agriculture financing considering that, agriculture is regarded as the foundation of the Ghanaian economy (MoFA, 2020), and needs special attention. MoFA (2020) suggests that, unlike the other sectors of the economy, the agriculture sector should be highly considered and given special attention. This consideration should reflect in the Bank of Ghana's policies established for commercial banks.

10.4.3 Discussion of the Results

The study found using the in-depth interview that the application of loan authorization procedures is one of the factors commonly used by commercial banks in agriculture finance. The finding confirms Luvsannyam et al. (2021) who posited that authorization procedures should be followed by banks before granting loans to borrowers to minimize risk exposure. It also agrees with the credit referencing theory stating all information about borrowers should be appraised before an authorization is granted. Our finding also supports the views of BoG (2020) which indicated that there must be a regular review of all authorization processes such as credit limit of borrowers' repayment capacity of borrowers, borrowers' business progress before granting credits to borrowers. Furthermore, this finding is supported by Mokatsanyane et al. (2017) who indicated that there should be, securitization of borrowers, before granting credit as a result of the high risk associated with the agriculture sector. Thus, authorization process is essential in credit risk management as Afriyie et al. (2018) concluded lending decisions that are not properly approved, appraised, or authorized most likely result in a high default rate and high credit risk exposure in commercial banks.

Also, the document analysis of Borrowers and Lenders Act, 2008 (Act 773) indicated that lenders do not have an effective way of searching for charges on properties that have been offered for credit to assure whether the charged property used as collateral is genuinely the bonafide property of borrowers as a result of endless litigations. The findings conform to the views of Dlugosch et al. (2018) who argued in a study conducted in Kenya that loan losses would be minimized particularly by commercial banks through insurance and traceable collaterals. This agrees with the liquidity theory as loan insurance in agricultural lending reduces losses and increases the liquidity position of commercial banks. Also, the BoG policy requirement that banks must reach a solid agreement with borrowers and strictly apply it to minimize credit losses was confirmed.

The interview results indicate that commercial banks have implementation challenges in agriculture financing as a result of the high risk associated with the sector. These include poor yield; lack of storage facilities and a ready market; inadequate agricultural infrastructure such as irrigation processes and processing facilities; improper bookkeeping; diversion of funds, difficulty in monitoring; and importation of cheaper agricultural produce. These results confirm the findings of Pham (2021), Luvsannyam et al. (2021) and the Ministry of Foods and Agriculture (MoFA, 2021) report of Ghana. This finding has some implications on the adverse selection theory propounded by Stglitz and Weiss (1981) which suggests that, lenders cannot differentiate between the appetite of borrowers and risk levels and that credit contracts are subject to limitations. It implies that where the debt responsibilities of agricultural borrowers are more than their cash flow they cannot repay loans with interest out of their pockets (Ahlin & Waters, 2016). Put in another way, borrowers only have the responsibility to pay loans with interest from the income generated from their agribusiness and not from their personal savings. This assertion suggests that borrowers will pay back loans if they have the means to do so and therefore the debt responsibility of borrowers does not extend to those who willingly default (Guttman, 2008). The limited liability of borrowers signals commercial banks to critically conduct due diligence works before advancing credits to agricultural borrowers in the agriculture value chain financing to avoid bearing all the credit risk associated with agricultural lending.

It was also found that most agricultural borrowers do not have all the requirements for an agricultural credit facility. This confirms the finding of Owusu-Boafo et al. (2020) on the inability of agriculturalists to meet borrowing requirements. It was also revealed that some of the commercial banks do not have dedicated and specialized units or officers purposely in charge of agriculture financing even though agricultural financing is a specialized area that must be given the needed attention. There is not much attention given to the agriculture sector despite the importance of the sector and its associated high risk. The interview results from both farmers and credit officers also confirmed a lack of insurance schemes in the agriculture sector, unlike the other sectors such as service and manufacturing which are heavenly supported with adequate larger agricultural credit share. This does not conform to Acharya et al. (2021) who found that insurance schemes for agricultural activities are very crucial in mitigating credit risk associated with agricultural lending in the United States. Even though the farmhouses are insured, the animals, farms, and the produces are not insured, making it challenging for commercial banks to confidently finance agriculture, as the risk involved is very high. The agriculture insurance would make monitoring and supervision difficult for the Banking Supervision Department of the Bank of Ghana to ensure compliance and minimize possible credit losses and hence reduce agricultural credit supply.

10.5 Conclusions and Recommendations

This study evaluated the effectiveness of the implementation of credit risk management policies by Ghanaian commercial banks in agricultural financing through the use of document analysis and interviews. According to the findings of this study, commercial banks are confronted with several challenges in the implementation of CRMPs as identified in the interview with farmers, credit, and policy documents. These challenges include improper and poor bookkeeping by agriculture borrowers; lack of insurance for agricultural produces; lack of collateral by borrowers; lack of collateral; poor yield leading to inadequate cash flow; underdeveloped export market; and inadequate attention on the agriculture sector financing. This study further revealed the lack of specialized units comprising of technical people with an agricultural background as a challenge in most commercial banks. Similarly, the review of policy documents indicated a lack of specific or specialized policy on agriculture financing established by the Bank of Ghana for commercial banks. We also found that most borrowers do not register their information with credit referencing bureaus institutions in Ghana. Also noteworthy is the fact that some borrowers divert the credits which have been secured for agricultural purposes into different areas not required and begin to avoid credit officers and therefore making supervision and monitoring difficult.

We recommend that agriculturalists must keep proper books of accounts to attract agricultural credits as indicated by Owusu-Boafo et al. (2020). Moreover, insurance schemes should be established and irrigation facilities provided for agricultural activities to mitigate losses and improve loan repayment. Further, proper and regular monitoring is recommended to prevent diversion of loans into other is not required. The agricultural sector should be supported with refined policy documents and implementation documents against the reality of borrowers' inability to honor loan contracts. By the new policy, BoG can specify concessions such as interest rate rebates, tax holidays (in consultation with Ghana Revenue Authority), for commercial banks who grant credits to borrowers in the agriculture value chain. Further, BoG must effectively and regularly supervise and monitor commercial banks; establish more credit guarantee schemes and incentive-based risk-sharing systems for small and medium agriculture enterprises; and establish more robust credit referencing bureau institutions such that all borrowers' credit history would be captured which would be assessed by commercial banks before granting loans for agriculture purposes to reduce credit risk.

Considering that a qualitative approach with a focus on CRMPs of the BoG was adopted in this chapter, an application of quantitative approach using questionnaires to elicit information on CRMPs of the commercial banks such as loan appraisal process, authorization procedures, loan approval processes, credit limit review, and other policies will shed more light on the subject matter. Be that it may, this study provides a framework for building the capacity of practitioners (the banking staff) and equipping them with empirically informed practices to mitigate credit risks associated with agricultural lending.

References

Abbas, S., Haider, S., Zainab, F., Hassan, H., & Fazal, A. (2018). Why the remodeling of risk management Practices in banking is required? Evidence from Pakistan. *International Journal of Scientific & Engineering Research, 9*(2), 686–691.

Abdelaziz, H., Rim, B., & Helmi, H. (2020). The interactional relationships between credit risk, liquidity risk, and bank profitability in the MENA region. *Global Business Review, 79*(5), 1–23. https://doi.org/10.1177/0972150919879304

Abu-Bakar et al., (2020). Board monitoring and covenant restrictiveness in private debt contracts during the global financial crisis. *Accounting & Finance, 60*(1), 661–692. https://doi.org/10.1111/acfi.12357

Acharya, V. V., Almeida, H., Ippolito, F., & Perez-Orive, A. (2021). Credit lines and the liquidity insurance channel. *Journal of Money, Credit and Banking, 53*(5), 901–938.

ADB. (2020). *2020 financial report*. GCB Bank.

Adusei, C. (2018). Determinants of non-performing loans in the banking sector of Ghana between 1998 and 2013. *Asian Development Policy Review, 6*(3), 142–154.

Afande, F. O. (2014). Credit risk management practices of commercial banks in Kenya. *European Journal of Business and Management, 6*(34), 21–62.

Afriyie, H. O., & Akotey, J. O. (2018). Credit risk management and profitability of selected rural banks in Ghana. *Catholic University College of Ghana, 7*(4), 176–181.

Ahlin, C., & Waters, B. (2016). Dynamic microlending under adverse selection: Can it rival group lending? *Journal of Development Economics, 121*, 237–257.

Amadhila, E., & Ikhide, S. (2016). Constraints to financing agriculture in Namibia. *African Review of Economics and Finance, 8*(2), 82–112.

Amedi, M., Dumayiri, M., & Mohammed, A. S. (2019). Loan repayment and its implication on agricultural financing in Ghana—The case of MiDA agriculture program. *International Journal of Agricultural Management and Development (IJAMAD), 9*(4), 391–408. https://doi.org/10.22004/ag.econ.301191

Apanga, M. A., Appiah, K. O., & Arthur, J. (2016). Credit risk management of Ghanaian listed banks. *International Journal of Law and Management, 58*(2), 178.

Ascui., F., & Conjoianu., T., F. (2019). Implementing natural capital credit risk assessment in agricultural lending. *Business Strategy and Environment*, *28*(6), 1234–1249.

Asiama, R. K., & Amoah, A. (2019). Non-performing loans and monetary policy dynamics in Ghana. *African Journal of Economic and Management Studies, 10*(2), 169–184.

Basel. (2015). Guidelines, corporate governance principles for banks. *Bank for international settlements.*

Boateng. K. (2019). Credit risk management and performance of banks in Ghana: The 'Camels' rating model approach. *International Journal of Business and Management Invention*, *8*(2), 41–48.

BoG. (2010). *Banking sector reports, 2010.* Bank of Ghana.

BoG. (2017). *Banking sector report 2017*. Bank of Ghana.

BoG. (2018). *Banking sector report 2018.* Bank of Ghana.

BoG. (2019). *Explanatory notes to the fit, and proper person directive for banks, savings and loans Companies, Finance Houses and Financial Holding Companies*. Bank of Ghana.

BoG. (2020). *Banking sector report. Collateral Registry. All you need to know.* Bank of Ghana.

BoG. (2021). *Banking sector reports, January 2021.* Bank of Ghana.

Braun, V., & Clarke, V. (2006). Using thematic analysis in psychology. *Qualitative Research in Psychology, 3*(2), 77–101.

Brown, M., Jappelli, T., & Pagano, M. (2009). Information sharing and credit: Firm-level evidence from transition countries. *Journal of Financial Intermediation, 18*(2), 151–172.

Caplinska, A., & Tvaronavičienė, M. (2020). Creditworthiness place in credit theory and methods of its evaluation. *Entrepreneurship and Sustainability Issues, 7*(3), 2542–2555.

Chowdhury, M., & Alam, Z. (2020). The impact of ethics on organizational performance: The case of private commercial banks of Bangladesh. *The USV Annals of Economics and Public Administration, 20* (2 (32)), 58–71.

Crawford, G. S., Pavanini, N., & Schivardi, F. (2018). Asymmetric information and imperfect competition in lending markets. *American Economic Review, 108*(7), 1659–1701.

Dlugosch, T. J., Klinger, B., Frese, M., & Klehe, U. C. (2018). Personality-based selection of entrepreneurial borrowers to reduce credit risk: Two studies on prediction models in low-and high-stakes settings in developing countries. *Journal of Organizational Behavior, 39*(5), 612–628.

Flatnes, J. E., & Carter, M. R. (2019). A little skin in the game: Reducing moral hazard in joint liability lending through a mandatory collateral requirement. *Journal of Economic Behavior & Organization, 164*, 199–214.

Freimer, M., & Gordon, M. J. (1965). Why do bankers ration credit? *The Quarterly Journal of Economics, 79*(3), 397–416.

GCB. (2020). *Financial report 2020. Annual report*. GCB Bank.

Glenn, A. B. (2009). Document analysis as a qualitative research method. *Qualitative Research Journal, 9*(2), 27–40. https://doi.org/10.3316/QRJ0902027

Gouri, K., & Mahajan, V. (2017). Different models of financing small farmers' agricultural value chains In *Financing agriculture value chains in India* (pp. 33–53). Springer. https://doi.org/10.1007/978-981-10-5957-5

GPIR. (2016). *Using the 6th Ghana Living Standards Survey*. Ghana

Guttman, J. M. (2008). Assortative matching, adverse selection, and group lending. *Journal of Development Economics, 87*(1), 51–56.

IMF. (2020). *Financial sector assessment program update. Basel core principles for effective banking supervision assessment of compliance international monetary fund*. Retrieved from Washington, D.C.

Kessey, D. (2015). Assessing credit risk management practices in the banking industry of Ghana: Processes and challenges. *Global Journal of Management and Business Research, 15*(6), 1–10.

Kitonga, P. (2017). Determinants of staff competence on the effectiveness of debt collection in commercial banks. *International Journal of Finance and Accounting, 2*(4), 24–45.

Kusi, B., Agbloyor, E., Fiador, V., & Osei, K. (2016). Credit referencing bureaus and bank credit risk: Evidence from Ghana. Evidence from Ghana. *Africa Finance Journal*, *18*(2). https://doi.org/10.10520/EJC199614

Lagat, F. K., Mugo, R., & Otuya, R. (2018). Effect of credit risk management practices on lending portfolio among savings and credit cooperatives in Kenya. *European Journal of Business and Management*, *5*(19), 93–105.

Luvsannyam, D., Minjuur, E., Lkhagvadorj, D., & Bekhbat, E. (2021). Non-performing loan recovery: The case of Mongolia. *Journal of Applied Finance and Banking*, *11*(1), 81–97.

Madugu, A. H., Ibrahim, M., & Amoah, J. O. (2020). Differential effects of credit risk and capital adequacy ratio on the profitability of the domestic banking sector in Ghana. *Transnational Corporations Review*, *12*(1), 37–52.

Maloba, M., & Alhassan, A. L. (2019). Determinants of agri-lending in Kenya. *Agricultural Finance Review*, *79*(5), 598–613.

Misra, M. (2021). Commercial micro-credit, neo-liberal agriculture and small-holder indebtedness: Three Bangladesh villages. *Journal of Contemporary Asia*, *51*(2), 330–350.

MoFA. (2020). *Report on sector performance*. Ministry of Finance.

MoFA. (2021). *Report on sector performance*. Ministry of Finance.

Mohd-Amin, S. I., & Abdul-Rahman, A. (2020). The role of regulation in Banking: Liquidity risk perspective. *Iranian Journal of Management Studies*, *13*(3), 391–412.

Mokatsanyane, D., Muzindutsi, P., & Viljoen, D. (2017). Credit risk and securitisation in the South African banking sector. *ACTA UNIVERSITATIS DANUBIUS*, *13*(2), 102–121.

Molefe, B., & Muzindutsi, P.-F. (2016). Effect of capital and liquidity management on profitability of major South African Banks. In *Proceedings of the 28th Annual Conference of the Southern African Institute of Management Scientists* (pp. 686–696), Pretoria, South-Africa.

Monye, F. N., Nwafor, N., & Mukoro, B. (2020). Appraisal of the Nigerian credit reporting act. *Journal of African Law*, *64*(2), 229–243.

Moss., D., B., & Suh., D., H., (2020). Effect of compliance cost on the supply of bank credit to agriculture: A differential approach. *American Journal of Agricultural Economics*, *102*(2), 713–726.

Nguyen, H. H. (2020). Individual customers' access to credits at commercial banks in Viet Nam: The case of Tra Vinh Province. *The Journal of Asian Finance, Economics, and Business*, *7*(9), 371–376.

Nwude, E., K., & Okeke, C. (2018). Impact of credit risk management on the performance of selected Nigerian banks. *International Journal of Economics and Financial*, *8*(2), 243–254.

Obuobi, B., Nketiah, E., Awuah, F., & Amadi, A. G. (2019). Recapitalization of banks: Analysis of the Ghana banking industry. *Open Journal of Business and Management, 8*(01), 78.

Odonkor, A. A. (2018). An assessment of credit risk management practices of Adansi Rural Bank Limited. *International Journal of Economics and Finance, 10*(11), 110–136. https://doi.org/10.2139/ssrn.3252463

O'Har, J. P., Senesi, C. W., & Molenaar, K. R. (2017). Development of a risk register spreadsheet tool for enterprise-and program-level risk management. *Transportation Research Record, 2604*(1), 19–27. https://doi.org/10.3141/2604-03

Ofori-Nyarko, E. (2017). The effect of supply chain management on firm performance: The case of selected commercial banks in Ghana. *Pentvars Business Journal*, *10*(2), 27–42.

Owusu-Boafo, R., Obeng, E., & Addo, J. Y. (2020). The relationship between credit risk management and the profitability of banks in Ghana. *Economic Studies & Analyses/Acta VSFS, 14*(2), 92–114.

Pham, H. N. (2021). How does internal control affect bank credit risk in Vietnam? A Bayesian analysis. *The Journal of Asian Finance, Economics, and Business, 8*(1), 873–880.

Rufo, M., & John, P. R. (2017). The effect of credit risk and capital adequacy on the profitability of rural banks in the Philippines. *Scientific Annals of Economics and Business, 64*(1). https://doi.org/10.1515/saeb-2017-0006

Saruni, L. A., & Koori, J. (2020). Credit information sharing and default rate of loans issued by commercial banks listed at the nairobi securities exchange. *Journal of Finance and Accounting*, *4*(5), 1–24.

Stiglitz, J. E., & Weiss, A. (1981). Credit rationing in markets with imperfect information. *American Economic Review, 71*(3), 393–410.

Udom, I. S., & Onyekachi, R. E. (2018). Effect of capital adequacy requirements on the profitability of commercial banks in Nigeria. *International Research Journal of Finance and Economics, 165*, 79–87.

Vallee, B., & Zeng, Y. (2019). Marketplace lending: A new banking paradigm? *The Review of Financial Studies, 32*(5), 1939–1982.

Wubin, S., Arthur, J., & Agyapong, E. K. (2020). *Financing SMEs in Ghana: Evidence of the optimal guarantee ratio.* Paper presented at the Conference on Interdisciplinary Business and Economics Research, Sydney, Australia.

CHAPTER 11

Evaluation of Lending Methodologies Used by Ghanaian Banks to Extend Credit to SMEs

Kofi Nyarko Gyimah, Joseph Olorunfemi Akande, and Paul-Francois Muzindutsi

11.1 Introduction

Banks have found credit assessment very challenging for a long time. Lots of literature abound on the lending methodologies applied by banks and other lending institutions in assessing SME loan applications (Altman et al., 2018; Yoshino & Taghizadeh-Hesary, 2018). Some of this literature has provided empirical evidence on these methodologies either in isolation (Berger & Black, 2011) or have sought to find complementarity among these methodologies (Uchida, 2011). Banks collect information from their clients in making lending decisions. The information can be

K. N. Gyimah · P.-F. Muzindutsi (✉)
School of Accounting, Economics and Finance, University of KwaZulu-Natal, KwaZulu-Natal, South Africa
e-mail: MuzindutsiP@ukzn.ac.za

K. N. Gyimah
e-mail: 217079806@stu.ukzn.ac.za; kngyimah@pentvars.edu.gh

J. A. Peprah et al. (eds.), *Financial Sector Development in Ghana*, Palgrave Macmillan Studies in Banking and Financial Institutions,
https://doi.org/10.1007/978-3-031-09345-6_11

either qualitative or quantitative. Information is very vital to banks as it indicates the borrower's behaviour towards handling finances. Banks can determine, from the information collected, a borrowers' ability to repay the loan with the interest charged. Banks in doing so go through a lot of stress by adopting different methodologies to assess the creditworthiness of their client (Aga & Reilly, 2011). Effective assessment of credit applicants is seen as a very essential component of the overall process to reduce credit risk. It is crucial not just for credit-related banking organizations, but also to the financial sector as a whole (Win, 2018). There are no unique assessment methodologies that seek to prioritize lending to SMEs by taking into account their peculiar challenges as such SMEs continue to experience higher credit constraints as compared to large firms (Poku & Frimpong, 2009; World Bank, 2019). This has been attributed to the lack of appropriate methods in assessing SMEs' creditworthiness (Yong et al., 2021). This has necessitated a lot of research by both financial practitioners and academic scholars in search of an appropriate methodology using different theoretical underpinnings and research methodologies.

One method that has been in use for centuries according to literature is the 5C's of lending and comprises of character, capacity, condition, collateral and capital (Dheeraj, 2017). This has been the principle frequently used by bank credit officer in evaluating credit proposals. It is a basic standard operating procedure banks use to identify customers who are eligible and trustworthy for credit (Manurung, 2019). Currently, the banking system has seen a growth in the amount of numerical information about borrower's creditworthiness resulting from the improvement in technology. As a result, some data which hitherto resided with banks can now be moved outside the bank and shared with other banks through credit reference bureaus. A bank's decision to use or rely on soft or hard information (or both) is dependent on the availability of the information and the advantages that such information will bring to the bank.

Despite this acknowledgement of the use of lending methodology, the central question that remains to be answered is whether the methodologies applied are also applicable to Ghanaian SMEs. If so, is methodology

J. O. Akande
Department of Accounting Science, Walter Sisulu University, Mthatha, Eastern Cape, South Africa
e-mail: jakande@wsu.ac.za

applied causing the difficulty experienced by SMEs in accessing credit from banks? These are the major questions to be answered by the current study. The contribution of the study to the body of knowledge is premised on the fact that previous study in Ghana on the use of lending methodologies by banks has focused on large firms and less on SMEs whose activities are mostly informal and lack the necessary credit records and financial history (Bank of Ghana, 2019). Existing studies in the field have also relied on bank data (i.e., credit files) to determine the effect of lending methodologies on access to credit (PwC, 2019). This study brings a new dimension to the provision of financial services by identifying lending methodology applied by Ghanaian banks from the firms' (SMEs' in this study's context) viewpoint. With this approach, this study determines the dominant lending methodologies implemented by Ghanaian banks to extend credit to SMEs. As a result, the research question as to the methods undertaken by the banks and the basis for assessing SME credit proposals is answered.

The remainder of the chapter is outlined as follows. Section 11.2 covers the literature review and outlines the analytical steps of the study. Section 11.3 provides information as to the research approach describing the data set and the procedure used in analysing the data. Sections 11.4 and 11.5 report on the results and discuss the major findings respectively. Section 11.6 provides a study, conclusion and recommendations based on the results of the study.

11.2 Literature Review on Lending Methodologies

To get a better understanding of the debate on lending methodologies used by banks in their dealings with SME firms, it is important to review some of the several studies that have been carried out globally on the lending methodology adopted by banks. Banks in dealings with SMEs which are perceived to be informationally opaque do collect data either on their own are aided by the central authority through credit information sharing to reduce this information problem and to ensure efficient allocation of resources. Soares et al. (2011) posit that such information can be either qualitative or quantitative. This information so collected aids banks in making credit decisions and is based on four main types of information (Hasanaj & Kuqi, 2019; Mayo, 2012): (i) information collected purely through a bank-SME relationship that is commercial and mostly qualitative; (ii) information of financial nature taking from financial records; (iii) information relating to the owners and managers of the firm and (iv) information which seeks to mitigate the risk of non-payment (collateral or guarantees in fulfilment of Basel 1 and 11 accords).

Banks can determine, from the information collected, whether a borrower can repay the loan with the interest charged. Banks in doing so go through a lot of stress by adopting different methodologies to assess the creditworthiness of their client (Aga & Reilly, 2011). The 5Cs is an acronym derived from the first letters of the specific areas considered when assessing loan applicants (Manurung, 2019). It represents the qualitative aspect of the lending methodologies and represents Character, Capacity, Capital, Collateral and Condition (Manurung, 2019). Character assesses the borrower's willingness to repay the amount of money requested and the accompanying interest and capacity is used to access the financial situation of the borrower and the ability to repay the loan on time (Chaibi & Ftiti, 2015). Capital refers to the borrower's ability to generate enough revenue from its business to repay the amount borrowed; collateral represents the borrower's ability to cover the loan with some form of security or guarantees; and condition is used to determine the macroeconomic circumstance prevailing that might render the borrower unable to meet its repayment obligation on time (Quartey et al., 2017).

11.2.1 The 5Cs Principle of Credit Assessment

The 5C's method of assessing the creditworthiness of applicants is based on the predictive modelling methodology. Loan applicants are assessed based on their character, capacity, capital, collateral and condition. Character is used here to refer to firm profile characteristics such as firm size, firm age (number of years in operation), ownership structure, location and business type all affect access to credit (Cowling et al., 2012). To evaluate capacity, banks rely on the analysis of financial statements to predict a borrower's repayment capacity (Kessey, 2015). The capital represents the size of the owners' capital introduced into the business and is a strong determinant of access to finance (Quartey et al., 2017). Collateral is the guarantee that the prospective borrower uses to secure the credit be it movable or unmovable. The condition represents the prevailing economic environment. This determines whether the existing economic condition supports the effective performance and survival of the business to generate enough returns to be able to repay the loan (Adeyeye et al., 2016). The 5Cs method has served banks and SME firms well, however, it has become evident over some few decades (especially after the emergence of the credit crunch) that this method was not robust enough to provide adequate information on which banks could use to make a good judgement of borrowers (Chaibi & Ftiti, 2015). Banks have faced significant obstacles using the 5Cs in the credit decision process mainly because prospective borrowers do not prepare financial statements making it difficult for credit officers to adequately measure for example capacity by using the accounting data to measure capacity variables let alone calculate the use of capital to determine the prospective applicant ability to pay and effective use of the loan granted (Manurung, 2019). This has necessitated a lot of research by both economic and academic scholars in search of an appropriate methodology using different theoretical underpinnings and methodologies.

11.2.2 Relationship and Transaction-Based Lending Methodologies

For several years researchers and policymakers have sought to investigate the problem of asymmetric information all aimed at reducing its effect on lending to SMEs. The approach has been to determine the appropriate lending methodology rather than focus on individual signing of commercial contracts (Carbo-Valverde et al., 2009). One methodology that has been identified in the literature as capable of reducing the effect is relationship lending (Petersen & Rajan, 1994). Relationship lending is seen to play an important role in financing SME firms within the context of information asymmetry in that it has a screening mechanism and a monitoring strategy capable of reducing the opacity of information associated with SME firms (Berger & Udell, 2006). Relationship lending is a technique designed to address the problem of information asymmetry between lenders and borrowers (Petersen & Rajan, 1995). The value of relationship lending becomes manifest as SMEs are seen to be informationally opaque coupled with the lack of credit history, the impossibility of credibly verifying the quality of the loan application as well as the lack of separation between ownership of the firm and management.

Transaction-based lending methodologies is that which are primarily based on hard financial information. Regarding small businesses, a firms' access to credit is premised on how much information is available to the financial market. That is small firms that are seen to be informationally transparent (i.e., those that can produce formalized records have a higher probability of having their loan request approved) (Degryse et al., 2017). Comeig et al. (2015), researching the Spanish SME credit market using a data set of 734 SME firms, found evidence that SME access to finance is enhanced by the extent of knowledge that lending institutions possess about loan applicants.

One major feature of transactional lending methodology is that information about clients can be relatively verified easily, observed and transmitted through internal communication channels with the financial institution. It includes Collateral-Based Lending, Financial Statement Lending, Credit Scoring and Viability-Based Lending methodologies (Angori et al., 2019). These transactions lending methodologies are meant to reduce the informational opacity of firms that operate informally and have no credit history. Ferri et al. (2019) in their analysis of banking data collected from some European countries concluded that firms that are assessed using transactional lending methodologies have a higher probability of having their loan demand approved as against firms that are assessed using relationship lending methodologies.

Prior studies have shown that information asymmetry influences the lending behaviour of banks to SMEs (De Haas & Van Horen, 2013). However, the extent to which banks can mitigate the effect depends on the methodologies applied in reaching a lending decision (Beck et al., 2018; Li et al., 2019; Sette & Gobbi, 2015). The use of different methodologies in arriving at a lending decision has indeed been a prevailing phenomenon in the banking sector. Petersen and Rajan (1994), posit that transactional lending methodology has consistently been found to be suitable for large corporate lending as opposed to relationship lending methodology that is well appropriate for SME lending. Hauswald and Marquez (2006) found that transactional (hard) lending information and relationship (soft) lending information all play different but significant roles in the lending process. Bank mostly relies on one type of information for a particular credit product. However, Bartoli et al. (2013), had a contradictory view by stating that lending technologies can be applied at the same time as they complement each other. Investigating the possibility of banks whether big or small combining lending methodologies for financing SMEs, Bartoli et al. (2013) concluded that when both lending methodologies are used together it gives more advantages to small businesses than using each technique exclusively. All these studies (Bartoli et al., 2013; Beck et al., 2018; De Haas & Van Horen, 2013; Li et al.,

2019; Sette & Gobbi, 2015) have relied on panel data from banks (credit files) in determining the effect of the lending methodologies used by banks, but the focus has mostly been on large firms. This current study fills the gap by relying on primary data from SME firms in Accra to evaluate bank lending methodologies to extend credit. The focus is primarily to identify the key methodologies used by lending institutions in assessing SME loan proposals.

11.3 Methodology

This study employed a quantitative technique in data gathering. The specific research approaches adopted were the correlational technique used to test some associations and the phenomenological approach utilized to explore the opinions of experts. The study was also a cross-sectional survey as it gathered and analysed data at a fixed period. The population for the study was SMEs in Accra that had acquired credit facilities from one or more banks. The simple random sampling method was used to select 1,061 SMEs from a target population of 8055 SMEs in Accra. The simple random sampling method is a probability sampling technique that gives all population members the same chance of being selected into the sample (Allwood, 2012). The simple random sampling approach determines the most representative sample because it accords population members the same probability of being selected into the sample.

Exploratory analysis was conducted before hypothesis testing. The standard procedure applied in literature (Na-Nan & Wongsuwan, 2020; Palmgren et al., 2018) was used to assess the psychometric indicators of the scale. The specific procedure used is the confirmatory factor analysis (CFA), and was used to estimate validity measures (i.e., average variance extracted (AVE) and maximum squared variance (MSV)). The maximum likelihood method was chosen as a means of factor extraction based on recommendations from Na-Nan and Wongsuwan (2020). The Cronbach's alpha coefficient (α) was computed as an indicator of scale reliability (Asiamah et al., 2018). Continuous variables including credit referencing were summarized using the mean and its standard deviation. Factor analysis (i.e., principal components extraction method) was then used to estimate relevant psychometric properties in an attempt to assess the reliability and validity of the scale. Cronbach's alpha coefficient was estimated to assess the internal consistency of the scales in harmony with procedures adopted from Kelava (2016) who have indicated that factor analysis is the ultimate foundation for understanding the factor structure of measurement scales. Subsequently, confirmatory factor analysis was used to estimate construct validity indicators, namely average variance extracted (AVE) and Mean Shared Variance (MSV). To test this hypothesis, the one-sample t-test, with a test value equals to the composite median, was used to test the hypothesis.

Although there are other methods such as SEM which is seen to be a more sophisticated and robust data analysis technique than multiple regression, researchers (Amorim et al., 2010; Tarka, 2018) have concluded that SEM is unnecessary if simple linear regression or Multiple Linear Regression (MLR) may be employed effectively, because there is no difference between estimates from SEM and regression in this circumstance. Tarka (2018) reasoned that using SEM in a situation where regression may be utilized could be seen as a beginner research technique. As a result, utilizing SEM in a circumstance where regression might be utilized can tarnish a study's reputation.

11.4 Data Analysis

The objective of this study was to examine the lending methodologies used by banks to evaluate SME credit proposals. The participants in this study were SMEs (represented by CEOs and managers) and credit risk officers of banks in Accra. SMEs provided data for the quantitative facet of this study. Of the 1,061 questionnaires administered to the SMEs, 899 were completed and returned by participants. Thirty (30) questionnaires were not completed at all and were therefore removed. For 21 of the questionnaires returned, participants did not respond to questions or items of a whole scale (e.g., lending methodologies). According to Garson (2012), questionnaires containing missing data for a whole scale should be discarded. As such, those 21 questionnaires were also dropped. Therefore, 80% of the questionnaires administered were analysed. Table 11.1 shows summary statistics of SMEs owners' demographic variables as well as basic business variables. Table 11.1 shows that about 65% ($n = 552$) of the participants were men whereas about 35% ($n = 296$) were women. Thus, the majority of participants in the study were men. This difference in the proportions of male and female participants implies that the findings of this study may not be generalized to the general population and could better represent the opinions of men. It is also evident that the largest proportion of participants were those aged 36–45. This result suggests that all the age groups were represented in the study, which provides a basis for generalizing findings to all age groups (Garson, 2012).

In terms of education, most participants had a master's degree. It can be seen in Table 11.1 that all participants had at least a senior high school certificate, which means that the sample of participants had some formal

Table 11.1 Basic owner and business characteristics

Variable	*Level*	*Frequency*	*Percent*
Gender	Male	552	65
	Female	296	35
	Total	848	100
Age	26–35	96	11
	36–45	232	27
	46–55	324	38
	Over 55	196	23
	Total	848	100
Education	High school	152	18
	Diploma/Bachelor's degree	228	27
	Master's	468	55
	Total	848	100
Ownership	Sole ownership	536	63
	Partnership	172	20
	Co-operate	140	17
	Total	848	100
Business category	Wholesale/retail	208	25
	Manufacturing	196	23
	Agriculture	100	12
	Services	344	41
	Total	848	100
Business age	1–5 yrs	56	7
	6–10 yrs	232	27
	11–20	208	25
	Above 20	352	42
	Total	848	100
Bank experience	1–5 yrs	56	7
	6–10 yrs	200	24
	11–20	284	33
	Above 20	308	36
	Total	848	100

Source Survey data (2020)

education. According to some researchers (Asiamah et al., 2017; Creswell, 2003), participants with at least a basic education qualification have a better ability to reason logically and provide accurate responses, especially in English. Thus, the above educational profile of the sample implies that participants had the knowledge and ability to provide accurate responses. In terms of the ownership of the businesses, most of the SMEs were sole proprietorships. Thus, less than 40% of the sample represented corporate and partnership businesses. The large proportion of sole proprietorships involved in this study suggests that the findings of this study may apply more to sole proprietorships. It also supports the argument of Essel et al. (2019) that SMEs in Ghana are dominated by sole proprietorships.

Table 11.1 further shows that the largest proportion of businesses that took part in the survey were in the services sector category. It has been reported that the dominant businesses of the SMEs sector in Ghana are service-based firms, agricultural businesses and manufacturing companies (Essel et al., 2019). The above statistics on the sectoral distribution of the sample, thus, indicate that the key groups that characterize the SMEs sector in Ghana are represented in this study.

Concerning the number of years, the firms have been in business the largest proportion of SMEs was over 20 years old. These statistics connote that every SME that participated in this study had existed for at least one year, and at least 60% of the sample had existed for at least 6 years. This result implies that all businesses had ample experience in their respective sectors, which may have enabled them to provide experience-based responses. According to Allwood (2012), responses based on ample industry and professional experience are the ideal data for studies utilizing primary data. It is noticeable from Table 11.1 that the largest proportion of SMEs had been banking for over 20 years. Needleless to say that participants of this study were required to provide responses regarding their borrowing activities and experiences with banks. The accuracy of the information provided by participants, therefore, depended on how long these participants had banked with their respective banks. The foregoing statistics, thus, suggest that participants (SMEs) had an appreciable level of banking experience on which their responses were based.

Table 11.2 shows the financial characteristics of SMEs, which indicate that the SMEs had different asset capacities and financial capabilities. Moreover, the sample contained more medium-sized businesses as at least 70% of the sample had an asset of more than GH¢750,000, equivalent to $129,310 (Essel et al., 2019). If so, the findings of this study may better describe medium-sized enterprises. Though these amounts are relatively large and could imply that the SMEs had satisfactory sales performance within the financial year, they do not give information about whether they met or exceeded projected (targeted) sales and resulted in profitability. This shortcoming is addressed by the next attribute, the annual income of the SMEs. Given the above results, it is apparent that at least 50% of the SMEs made an annual income of GH¢10,000 (equivalent to $1,724). While these incomes look appreciable, it is not clear whether they met the business' expectations or targets. It can, however, be said that the SMEs had a positive net income over the period.

Table 11.3 shows the summary statistics for measures to access credit. It can be seen that more than 60% of SMEs had received at least GH¢10,001 (equivalent to $1724) and all SMEs had received at least GH¢2,000 (equivalent to $345). This reflects the fact that the sample included only businesses, which had acquired a credit facility at least

Table 11.2 Financial characteristics of SMEs businesses

Variable	*Group*	*Frequency*	*Percent*
Fixed assets	Less than ¢50,000	148	17
	¢500,000–¢750,000	156	18
	¢750,001–¢1,000,000	256	30
	¢1,000,001–¢1,500,000	264	31
	More than ¢1,500,000	24	3
	Total	848	100
Current asset	Less than ¢50,000	44	5
	¢500,000–¢750,000	128	15
	¢750,001–¢1,000,000	132	16
	¢1,000,001–¢1,500,000	276	33
	More than ¢1,500,000	268	32
	Total	848	100
Annual sales (2019)	Less than ¢50,000	296	35
	¢500,000–¢750,000	552	65
	Total	848	100
Net income (2019)	Less than ¢10,000	136	16
	¢10,000–¢20,000	216	25
	¢21,000–¢30,000	320	38
	¢31,000–¢40,000	176	21
	Total	848	100

Source Survey data (2021)

once. This result suggests that SMEs had experienced banking lending and knew about the conditions and methodologies applied in lending. This finding implies that all SMEs had accessed credit facilities from the banks and therefore acquired some knowledge about bank lending.

Table 11.4 shows a summary statistic on lending methodology and its indicators. It is evident that all items of the scale produced relatively large mean scores and therefore represent lending methodologies applied to the SMEs. The interpretation of the scale this way is consistent with Asiamah et al. (2018), who applied the same descriptive anchors to analyse the level of emotional intelligence in a sample and concluded that medians close to the mean indicate relative consistency between the measures of central tendency. The item with the largest mean score is "I was asked if I had a collateral" (Mean = 4.36; SD = 0.72), followed by "I was interviewed on the policy, philosophy, and performance of my business" (Mean = 4.35, SD = 0.73). This finding suggests that the provision of collateral and information on the policy, philosophy and performance of the business were outstanding requirements for acquiring credit facilities. The least mean scores are produced by "I was asked about my age" (Mean = 3.21; SD = 0.81) and "I was asked about my gender" (Mean = 3.15; SD = 0.84). That is to say that the provision of information on the borrower's demographic characteristics is less of lending criteria applied.

Table 11.3 Measures of access to credit facilities

Variable	*Group*	*Frequency*	*Percent*
Access to loans	Yes	848	100
	No	0	0
	Total	848	100
Frequency of loan access	Once in a while	236	28
	Once a year	504	59
	Several times in a year	108	13
	Total	848	100
Loan received (¢)	Less than ¢2,000	12	1
	¢2,000–5,000	48	6
	¢5,001–¢10,000	248	29
	¢10,001–¢15,000	496	58
	More than ¢15,000	44	5
	Total	848	100

Source Survey data (2021)

However, this information was still an aspect of the lending methodologies. Lending methodologies as a variable accounted for a mean score of about 81 (Mean = 80.82; SD = 12.08) out of a maximum expected mean score of 95. This average score produces about 85% of the maximum scale score, which is in excess of the 50th percentile of the variable "lending methodologies". It can be seen from Table 11.4 that the medians are close to the mean. The median of "lending methodology", for instance, is 81, which is close to the corresponding mean of 80.82.

Table 11.5 shows the summary statistics on credit referencing information. Based on Asiamah et al.'s (2018) interpretation disclosed above, it can be said that all items in Table 11.5 produced relatively large mean scores. Therefore, the items represent a significant level of access to credit referencing information. The largest mean score was produced by "The procedure for obtaining bank loan has been simplified by the introduction of the use of credit referencing information" (Mean = 4.43; SD = 0.64), followed by "Sharing of credit information has reduced loan processing costs and increased lending volume significantly" (Mean = 4.41; SD = 0.64). "Sharing of credit information has reduced interest rates charged on loans by commercial banks" (Mean = 4.18; SD = 0.71) and "Credit information sharing has effectively reduced non-performing loans on the institution's financial statement" (Mean = 4.12; SD = 0.79) produced the lowest mean scores, though these scores are close to 5. Credit referencing information (CRI) produced a mean score of about 34 (Mean = 34.40; SD = 4.52) out of a total expected mean score of 40. Thus, CRI accounted for about 86% of the total scale score, which is in excess of the 50th percentile of the variable "CRI". Further to this, the median of the variables in Table 11.5 are close to their corresponding means.

Table 11.6 shows results from an assessment of the distribution of the key variables. The distribution of the data is assessed with summary statistics and the Shapiro–Wilk's test. Shapiro–Wilk's test examines the univariate normality of the data and accompanies relevant graphs. More specifically, the focus is on descriptive statistics on the two main variables, namely lending methodologies, and credit risk information. Lending methodologies accounted for skewness and kurtosis values of −0.49 and −0.87 respectively. CRI also accounted for skewness and kurtosis values of −0.44 and −0.86 respectively. According to Garson (2012), skewness and kurtosis values falling between −3 and +3 are satisfactory and indicate that there are no significant outliers in the data. Each variable further accounted for a Shapiro–Wilk's statistic ≥ 0.8 ($p < 0.001$).

Table 11.4 Summary statistics on lending methodology and its indicators

Variable	*Min*	*Max*	*Mean*	*Median*	*SD*
I was asked if I had a collateral	3	5	4.36	5.00	0.72
I was asked to provide my balance sheet information	3	5	4.33	4.00	0.73
I was interviewed on the policy, philosophy, and performance of my business	3	5	4.35	5.00	0.73
I was asked to provide a copy of my business plan	3	5	4.32	4.00	0.75
I was asked to provide or discuss my financial performance target	3	5	4.32	4.00	0.74
I was asked about the historical records of my sales	3	5	4.35	4.00	0.71
I was asked about any debts I had	3	5	4.36	5.00	0.72
I was asked about the recognition of my business	1	4	3.18	3.00	0.85
Collateral and banking records	23	39	33.58	35.00	5.16
I was asked about my age	1	4	3.21	3.00	0.81
I was asked about my gender	1	4	3.15	3.00	0.84
I was asked if I was married	1	4	3.17	3.00	0.86
I was asked about my highest education	1	4	3.21	3.00	0.86
I was asked to indicate my experience in the business	3	5	4.35	5.00	0.73
The number of years the business had been in operation was ascertained	3	5	4.32	4.00	0.75
I was made to provide information on the ownership structure	2	5	4.29	4.00	0.80
I was asked about the size of my business	2	5	4.31	4.00	0.78
I was asked to describe the structure of the industry I operated in	2	5	4.28	4.00	0.81
I was asked to indicate the number of years I had banked with the bank	3	5	4.32	4.00	0.75
I was asked to indicate if I had received a loan from the bank in the past	3	5	4.29	4.00	0.76
I was asked to indicate how many times I had received a loan in the past from the bank	3	5	4.33	4.00	0.73
Personal and business characteristics	32	56	47.24	48.00	7.21
Lending methodology	55	95	80.82	81.00	12.08

Source Survey data (2021)

Table 11.5 Summary statistics on credit reference information and its indicators

Variable	*Min*	*Max*	*Mean*	*Median*	*SD*
The procedure for obtaining a bank loan has been simplified by the introduction of the use of credit referencing information	3	5	4.43	5.00	0.64
Banks are responsive to SME financing needs	3	5	4.20	4.00	0.70
Sharing of credit information has reduced loan processing costs and increased lending volume significantly	3	5	4.41	4.00	0.64
Sharing of credit information has reduced interest rates charged on loans by commercial banks	3	5	4.18	4.00	0.71
Credit information sharing from Credit Reference Bureaus rewards and promotes good credit track record for firms	2	5	4.39	5.00	0.70
Credit information sharing acts a deterrent to loan default	3	5	4.24	4.00	0.70
Credit information sharing has impacted positively on loan repayment	3	5	4.42	5.00	0.66
Credit information sharing has effectively reduced non-performing loans on the institution's financial statement	2	5	4.12	4.00	0.79
Credit risk information	24	40	34.40	36.00	4.52

Note SD—standard deviation
Source Survey data (2021)

Table 11.6 Tests of data distribution and normality

Variable	*Mean*	*SD*	*Skewness*	*Kurtosis*	*Shapiro Wilk's test*
Lending Methodologies	80.82	12.08	−0.492	−0.87	0.912
Credit risk information	34.4	4.52	−0.441	−0.86	0.895

$^{**}p < 0.001$
Source Survey data (2021)

11.4.1 Reliability and Validity Tests

Exploratory Factor Analysis (EFA) and Cronbach's alpha coefficient are used to assess the reliability and validity of the measurement scales used in this study. Table 11.7 shows factors loadings and key statistics resulting from an EFA, specifically the maximum likelihood method with ProMax rotation. Two factors were extracted from the EFA. The first 9 items constitute the first factor, with each accounting for factor loadings ≥ 0.5 (0.68–0.91). The second factor is made up of the remaining items (i.e., 10–20) that account for factor loadings ≥ 0.5 (0.57–0.98). The first nine items represent information on collateral and business records (profile). As such, it is called "Collateral and business records" (CBRs). The second factor comprises items describing the demographic attributes of owners and business characteristics. It is therefore called "Personal and business characteristics" (PBCs). CBRs produce a variance of about 67% (Eigenvalue = 13.4) whereas PBCs account for a variance of about 21.3% (Eigenvalue = 4.27). The total variance accounted for by the two factors is 88.3%, which represents a good model fit. In the context of the EFA, a valid (extracted) factor is a set of items that produce an Eigenvalue not less than 1 (Garson, 2012; Kelava, 2016). Figure 11.1 is a scree plot to depict the number of factors extracted.

Figure 11.1 shows a scree plot depicting the number of factors extracted. In this figure, only two factors (i.e., CBRs and PBCs) have Eigenvalues greater than 1 as recommended by Garson (2012).

Table 11.8 shows findings from the EFA for CRI with Promax rotation. Since the scale used to measure CRI is a unidimensional construct, a single-factor extraction method was specified. Thus, as Table 11.8 indicates, only one factor was extracted, with each item producing a factor loading ≥ 0.5 (0.63–0.97). The total variance produced by the factor is 66.3%, which represents a high model fit (Kelava, 2016). The EFA results above suggest that the scales used to measure lending methodologies and CRI were internally consistent. Even so, further analysis was conducted to affirm the reliability and validity of the two measures.

Table 11.9 shows the composite reliability, Cronbach's alpha and average variance extracted from the EFA. The extant literature indicates that Cronbach's alpha is a measure of internal consistency whereas composite reliability (CR) is an estimate of scale reliability (Asiamah et al., 2018; Kelava, 2016). Furthermore, Cronbach's α confirms satisfactory internal consistency at $\alpha \geq 0.7$. It can be seen that this condition is met for each domain as well as the whole construct in Table 11.9. Besides, construct validity is achieved at CR ≥ AVE (Average variance extracted) while discriminant validity is achieved at AVE ≥ 0.5 for lending methodologies (Asiamah et al., 2018; Kelava, 2016). CR, AVE and MSV values were not computed for CRI for a couple of reasons. These statistics are computed based on information from theoretical factors extracted in

Table 11.7 Factor loadings and variance of lending methodology from EFA

	Variance extracted	
Eigenvalue	13.4	4.27
Variance (%, total = 88.31)	66.98	21.34
Indicator	1	2
1. I was asked if I had a collateral	0.55	0.16
2. I was asked to provide my balance sheet information	0.91	0.14
3. I was interviewed on the policy, philosophy, and performance of my business	0.91	0.14
4. I was asked to provide a copy of my business plan	0.89	0.17
5. I was asked to provide or discuss my financial performance target	0.89	0.09
6. I was asked about the historical records of my sales	0.95	0.16
7. I was asked about any debts I had	0.92	0.16
8. I was asked about the recognition of my business	0.68	0.11
9. I was asked about my age	0.11	0.96
10. I was asked about my gender	0.13	0.89
11. I was asked if I was married	0.11	0.98
12. I was asked about my highest education	0.14	0.97
13. I was asked to indicate my experience in the business	0.09	0.98
14. The number of years the business had been in operation was ascertained	0.05	0.95
15. I was made to provide information on the ownership structure	0.10	0.89
16. I was asked about the size of my business	0.11	0.88
17. I was asked to describe the structure of the industry I operated in	0.08	0.91
18. I was asked to indicate the number of years I had banked with the bank	0.09	0.93
19. I was asked to indicate if I had received a loan from the bank in the past	0.12	0.89
20. I was asked to indicate how many times I had received a loan in the past from the bank	0.07	0.57

Source Survey data (2021)

EFA (Slocum-Gori & Zumbo, 2011). As indicated above, since a one-factor scale was obtained, there is no basis for AVE, MSV and CR to be computed. Secondly, Kelava (2016) indicated that internal consistency assessed with the Cronbach's alpha coefficient is enough for a unidimensional scale since convergent validity and discriminant validity for such a scale cannot be computed with a single source primary data. These

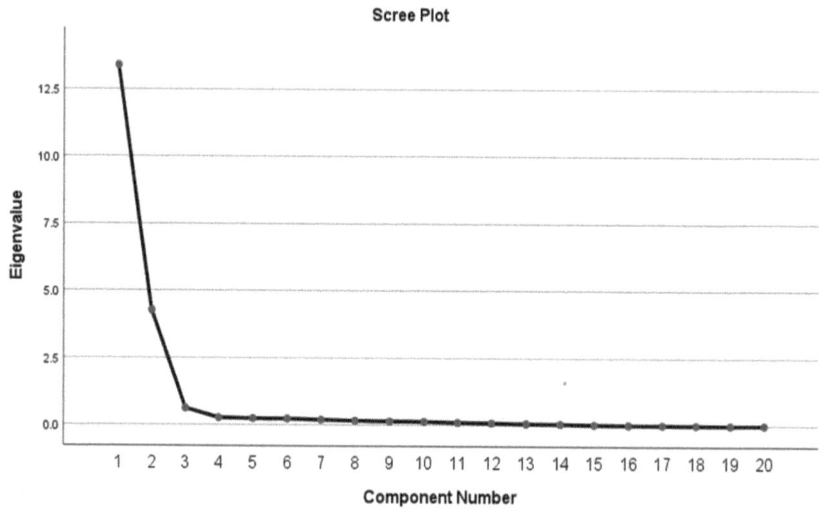

Fig. 11.1 PCA scree plot for factors of lending methodology (*Source* Survey data [2021])

Table 11.8 Factor loadings and variance of credit risk information from EFA

Statistics	*Factor loading*
Eigenvalue (5.34)	
Variance (%, total = 66.34)	
Indicator	
1. The procedure for obtaining a bank loan has been simplified by the introduction of the use of credit referencing information	0.97
2. Banks are responsive to SME financing needs	0.96
3. Sharing of credit information has reduced loan processing costs and increased lending volume significantly	0.96
4. Sharing of credit information has reduced interest rates charged on loans by commercial banks	0.51
5. Credit information sharing from Credit Reference Bureaus rewards and promotes good credit track record for firms	0.69
6. Credit information sharing acts as a deterrent to loan default	0.94
7. Credit information sharing has impacted positively on loan repayment	0.95
8. Credit information sharing has effectively reduced non-performing loans on the institution's financial statement	0.63

Source Survey data (2021)

Table 11.9 Psychometric properties of lending methodologies and credit risk information scales

Variable	*Factor*	*Cronbach's alpha*	*Composite reliability*	*Average variance extracted*	*Maximum shared variance*
Lending methodologies	Factor 1	0.937	0.954	0.621	0.212
	Factor 2	0.935	0.944	0.619	0.204
	Whole scale	0.966	–	–	–
Credit risk information	Whole scale	0.926	–	–	–

Source Survey data (2021)

thoughts explain why only Cronbach's alpha was computed and shown for CRI in Table 11.9.

11.4.2 Hypotheses Testing

This analysis builds on the EFA, which initially identified two domains of the lending methodologies (i.e., CBRs and PBCs). To address the objective, descriptive statistics and the one-sample t-test are used to examine the extent to which these methodologies were applied by the banks in lending to banks. The researcher draws on an existing approach (Altman & Royston, 2006) to examine the extent to which lending methodologies were used. In this vein, the median and average scores of lending methodologies and its two domains were estimated. According to Altman and Royston (2006), scores above the median represent a high extent of application of the lending methodology and vice versa. As such, the one-sample t-test was used to examine whether the average of the variable is significantly larger than its median.

The hypothesis (H1) tested is that the average score associated with lending methodologies is significantly higher than the median score of this variable. This hypothesis has two sub-hypotheses as follows:

H1a: the average score associated with CBRs is significantly higher than the median score of this variable.

H1b: the average score associated with PBCs is significantly higher than the median score of this variable.

If this hypothesis is confirmed, then the extent of use of lending methodologies can be said to be high. Table 11.10 shows descriptive statistics associated with the above hypotheses.

In Table 11.10, the median scores of CBRs, PBCs and lending methodologies are 24, 36 and 60 respectively. The mean score of CBRs (Mean = 33.58; SD = 5.16), PBCs (Mean = 47.24; SD = 7.21) and lending methodologies (Mean = 80.82; SD = 12.08) are seen to be larger than their medians. Figure 11.2 shows the proportion of the maximum value and median (see Table 11.10) that each variable accounted for as the average score. These proportions express the extent to which SMEs agreed that the lending methodologies were applied to them. It can be seen that the whole scale (i.e., lending methodology) produced an average score that represents 81% of the maximum score of 100 and 135% of the median. CBRs accounted for 84% of the maximum score and 140% of the median whereas PBCs accounted for 79% of the maximum score and 131% of the median. These proportions, thus, confirm that the average scores produced by the variables are larger than the median.

Table 11.11 shows a one-sample t-test used to test the above hypotheses. The test value of the *t*-test is the median of the variable involved. The test associated with CBRs is significant (t = 54.08, p = 0.000), which suggests that the mean score of CBRs is significantly larger than its median. The test is also significant for PBCs (t = 45.39, p = 0.000) and lending methodologies (t = 50.2, p = 0.000). The hypothesis and its sub-hypotheses are, therefore, confirmed. In other words, the extent of the application of the lending methodologies was high.

Table 11.10 Summary statistics on lending methodologies and its indicators

Variables	*Scale*			*Mean*	
	Minimum	*Maximum*	*Median*	*Statistic*	*SD*
Collateral and business records	8	40	24	33.58	5.16
Personal and business characteristics	12	60	36	47.24	7.21
Lending Methodologies	20	100	60	80.82	12.08

Source Survey data (2021)

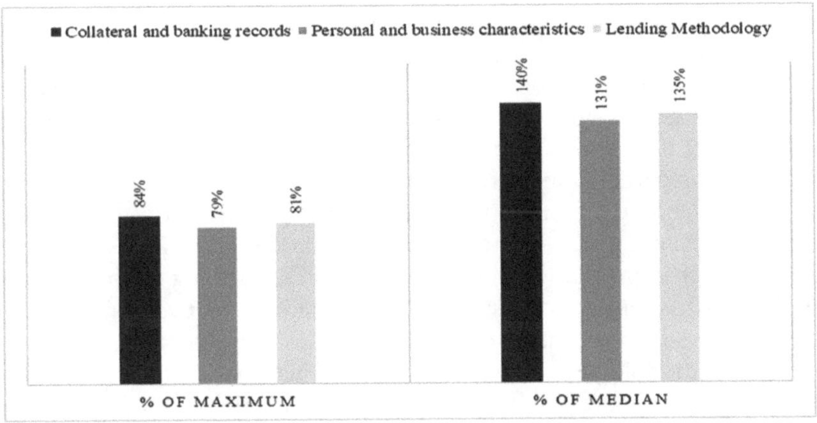

Fig. 11.2 Percent points indicating reported levels of application of lending methodologies (*Source* Survey data [2021])

Table 11.11 The one-sample *t*-test for a high level of application of lending methodologies

Variable	*Test value*[a]	t- *values*	*df*	*p-values*	*Mean difference*	*95% CI*	
						Lower	*Upper*
Collateral and business records	24	54.077	847	0.000	9.58	9.23	9.93
Personal and business characteristics	36	45.392	847	0.000	11.24	10.75	11.72
Lending methodology	60	50.199	847	0.000	20.82	20.00	21.63

[a]Value equal to the median in Table 11.10; CI—confidence interval
Source Survey data (2021)

11.5 Discussion of the Results

Our findings revealed that two categories of lending methodologies were used by the banks. The first category extracted from the exploratory factor analysis (principal components) is collateral and business records (CBRs), whereas personal and business characteristics (PBCs) constitute the second factor extracted. Commentaries provided in the literature suggest that factor extraction in exploratory factor analysis is based on the strength of the relationship between indicators of the factors (Kelava, 2016). The extraction of CBRs as the first factor thus connotes that the indicators of this factor are more strongly correlated and produce a larger variance, compared to indicators of the second factor extracted. In other words, the lending methodologies can be conceptualized construct of two related domains, which are the lending methodologies (i.e., CBRs and PBCs). More importantly, while EFA showed that lending methodologies used in Ghana had a factor structure characterized by two components, the analysis did not indicate the more important component of the lending methodologies. Even so, the current factor structure is important as it provides a theoretical basis for full-scale validation of scale measuring lending methodologies in Ghana or possibly other settings. This assertion is premised around the idea of Kelava (2016) and Asiamah et al. (2018) who postulate that the factor structure produced with EFA is a statistical blueprint or theoretical basis for conducting a confirmatory factor analysis and estimating reliability and validity statistics associated with a scale.

Further analysis of the objective indicates that CBRs (which is represented by manifest variables 1–8 in Table 11.7) a construct of the lending methodologies was perceived as the more important methodology as they provided a larger average score. In the context of this study, being the most important sub-methodology means that it is the most frequently used by banks in Ghana. This result is consistent with several studies and commentaries in the literature (Angori et al., 2019). The first example is Duarte et al. (2017) who reported that collateral-based lending is the most frequently used methodology in the banking sector due to its potential to protect financial institutions. Ferri et al. (2019) also acknowledge that collateral-based lending is the most dominant form of lending in the European region that protects both the creditor and borrower. Mills and McCarthy (2014) also revealed that lending methodologies asking for collateral security and information on borrowers are among the most reliable approaches in the world. They added that other methods such as PBCs are only supplementary. Finally, Ramalho et al. (2018) affirmed that the only sustainable approach to lending or the provision of credit for business expansion is the use of more secure methods such as collateral-based lending.

Data analysis also showed that both Collateral Based Records (CBRs) and Personal Business Characteristics (PBCs) methodologies are sufficiently used by the banks, though CBRs was more important and frequently used. This result came from the one-sample t-test, which confirms that the extent of use of both methodologies was high. This result is congruent with the reasoning of the Ramalho et al. (2018) that lending methodologies or criteria better work together as a complementary set. This is to say that the two methodologies play unique roles and that failing to apply both makes a lending process vulnerable to credit risks. Duarte et al. (2017) also found that all lending methodologies and criteria are considered important as they play unique roles to credit risk mitigation. For instance, if a borrower provides collateral security for a credit facility but does not provide information on his or her business, this may make it difficult for the bank to manage and track the collateral provided as well as monitor how well the individual is repaying the loan facility. This is rightly so because debt recovering, and loan repayment are assessed with the information provided by the borrower at the time of loan application.

This study's findings are consistent with both the current practice and theory of lending as described in the literature. With the EFA retaining all items with which lending methodologies were measured, this study supports the credit referencing information theory (Gehrig & Stenbacka, 2007), which describes a set of criteria that banks consider making lending decisions. This model assumes that the bank or creditor provides credit facilities after securing collateral security from the borrower as well as information that links the collateral to the borrower. While this paradigm is the most frequently used methodology (Ferri et al., 2019; Ramalho et al., 2018), it is different from other models used, particularly Islamic lending models that do not impose interests on loans (Mills & McCarthy, 2014). It is thus understandable that Ghana applies the traditional model because its banking sector is free of Islamic banks. Even so, Islamic lending methodologies do not overlook collaterals and information on borrowers (Ramalho et al., 2018), as a result of which such approaches can be consistent with the traditional approach. The ideal way to differentiate Islamic models from the traditional methodologies confirmed in this study is that the former is less stringent or more flexible to borrowers (Mills & McCarthy, 2014).

11.6 Conclusions and Recommendations

The purpose of this chapter was to evaluate the extent of use of lending methodologies by Ghanaian banks. Sampled banks employed two lending methodologies, namely CBRs and PBCs. The use of CBRs was focused on ensuring that SMEs provided the bank with collateral security in the form of personal assets (e.g., a car or house). This aspect of the methodology also emphasized the need for the SME to have banking experience and the ability to repay a loan facility. PBCs' utilization is aimed at linking a loan facility to the identity of the borrowing SME. Without this aspect of the methodology, the creditor or bank has no information with which to recover debts. Though CBRs was applied to a higher extent, both methodologies were used by the banks. It is therefore very necessary for SME owners to be conversant with the methodologies used by banks to assess loan proposals to increase the probability of getting a favourable decision. Banks should also develop their internal capabilities to effectively deal with SME credit proposals.

The objective of this study was to identify the lending methodologies used by banks to provide credit facilities to SMEs. With this objective, the researcher also sought to investigate the extent of the use of the lending methodologies. Generally, findings from this objective provide key lessons for enabling SMEs to better meet the lending methodologies in their effort to access credit facilities from banks. This is rightly so because knowing the specific lending methodologies and the extent to which they influence lending are a requirement for meeting these lending methodologies and knowing how to prepare to apply for credit facilities. Based on the findings of the objective, therefore, the specific recommendations to SMEs are as follows:

- Improving knowledge of lending methodologies used. The analysis of the objective reveals that two lending methodologies are used by banks in lending to SMEs. Going forward, SMEs ought to know these methodologies, their differences and how well they are currently prepared to meet them should they decide to apply for credit facilities.
- SME firms should strive to understand the extent of the application of lending methodologies. It is necessary to understand the extent of the application of the methodologies. The current study shows that CBRs is more frequently used, but both methodologies

influence lending decisions. By knowing the extent of the use of a methodology, the SME may decide to prioritize some elements in their effort to prepare to apply for credit facilities. For example, it may become obvious that collateral security and banking experience are the most important aspects of the methodology. If so, the SME may decide to first work towards building valuable collateral security and developing a banking relationship with an appropriate bank.

- SME firms should seek to align business operations with lending methodologies.

  Developing and strengthening its relationship with the bank sets the foundation for doing things right to improve creditworthiness. With a strong relationship with its bankers, the business would access relevant information for aligning operations with the lending methodologies or the entire lending process. SMEs can also adjust their expenditure to ensure that more revenues generated go to the bank or into securing valuable collateral security for borrowing purposes. This recommendation particularly applies to new SMEs that need loans to grow but are unable to secure credit facilities because of their low banking experience and lack of collateral security.

An important limitation of the chapter is the selection and use of SME firms ignoring large firms as well as the reliance only on samples selected from the Greater Accra region of Ghana. This may lead to participants' biases that may affect the reliability of the analysis as most of the firms are relatively young and may not have the track record to provide the full credit information sharing data relative to a large firm. Future research should consider including both small and large firms throughout the country in its data collection to deal with the limitation to facilitate generalization.

References

Adeyeye, P. O., Azeez, B. A., & Aluko, O. A. (2016). Determinants of small and medium scale enterprises financing by the banking sector in Nigeria: A macroeconomic perspective. *Investment Management and Financial Innovations, 13*, 170–175. https://doi.org/10.21511/imfi.13(1-1).2016.04

Aga, G. A., & Reilly, B. (2011). Access to credit and informality among micro and small enterprises in Ethiopia. *International Review of Applied Economics, 25*, 313–329.

Allwood, C. M. (2012). The distinction between qualitative and quantitative research methods is problematic. *Quality & Quantity. International Journal of Methodology, 46*(5), 1417–1429. https://doi.org/10.1007/s11135-011-9455-8

Altman, E. I., Esentato, M., & Sabato, G. (2018). Assessing the credit worthiness of Italian SMEs and mini-bond issuers. *Global Finance Journal, 43*. https://doi.org/10.1016/j.gfj.2018.09.003

Altman, D. G., & Royston, P. (2006). The cost of dichotomising continuous variables. *BMJ, 332*(7549). https://doi.org/10.1136/bmj.332.7549.1080

Amorim. L. D. A. F., Fiaccone, R. L., Santos, C. A. S. T., dos Santos, T. N., de Moraes, L. T. L. P., Oliveira, N. F., Barbosa, S. O., dos Santos, D. N., dos Santos, L. M., Matos, S. M. A., & Barreto, M. L. (2010). Structural equation modeling in epidemiology. *Cad. Saúde Pública, 26*(12): 2251–2262, dez.

Angori, G., Aristei, D., & Gallo, M. (2019). Lending technologies, banking relationships, and firms' access to credit in Italy: The role of firm size. *Applied Economics, 51*(58), 6139–6170. https://doi.org/10.1080/00036846.2019.1613503

Asiamah, N., Mensah, H. K., & Danquah, E. (2018). An assessment of the emotional intelligence of health workers: A scale validation approach. *Journal of Global Responsibility, 9*, 1–24.

Asiamah, N., Mensah, H. K., & Oteng-Abayie, E. F. (2017). General, target and accessible population: Demystifying the concepts for effective sampling. *The Qualitative Report, 22*(1), 1–14.

Bank of Ghana. (2019). *Financial stability reports*.

Bartoli, F., Ferri, G., Murro, P., & Rotondi, Z. (2013). SME financing and the choice of lending technology in Italy: Complementarity or substitutability? *Journal of Banking & Finance, 37*(12), 5476–5485.

Beck, T., Degryse, H., De Haas, R., & van Horen, N. (2018). When arm's length is too far: Relationship banking over the credit cycle. *Journal of Financial Economics, 127*(1), 174–196. https://doi.org/10.1016/j.jfineco.2017.11.007

Berger, A. N., & Black, L. K. (2011). Bank size, lending technologies, and small business finance. *Journal of Banking & Finance, 35*(3), 724–735. https://doi.org/10.1016/j.jbankfin.2010.09.004

Berger, A. N., & Udell, G. (2006). A more complete conceptual framework for SME finance. *Banking and Finance, 30*(11), 2945–2966.

Carbo-Valverde, S., Rodriguez-Fernandez, F., & Udell, G. F. (2009). Bank market power and SME financing constraints. *Review of Finance, 13*(2), 309–340.

Chaibi, H., & Ftiti, Z. (2015). Credit risk determinants: Evidence from a cross-country study. *Research in International Business and Finance, 33*, 1–16.

Comeig, I., Fernández-Blanco, M. O., & Ramírez, F. (2015). Information acquisition in SME's relationship lending and the cost of loans. *Journal of Business Research, 68*(7), 1650–1652. https://doi.org/10.1016/J.JBUSRES.2015.02.01

Cowling, M., Liu, W., & Ledger, A. (2012). Small business financing in the UK before and during the current financial crisis. *International Small Business Journal, 30*(7), 778–800.

Creswell, J. W. (2003). *Research design: Qualitative, quantitative, and mixed methods approaches* (2nd ed.). SAGE Publications.

Degryse, H., Matthews, K., & Zhao, T. (2017). Relationship banking and regional SME financing: The case of Wales. *International Journal of Banking, Accounting and Finance, 8*(1), 93–118.

De Haas, R., & Van Horen, N. (2013). Running for the exit? International bank lending during a financial crisis. *The Review of Financial Studies, 26*(1), 244–285. https://doi.org/10.1093/rfs/hhs113

Dheeraj, V. (2017). *Credit analysis: What credit analyst look for? 5Cs credit analysis*. Investment and Banking Training. JP Morgan Equity, IIT Delhi.

Duarte, F. D., Gama, A. P., & Esperanca, J. P. (2017). Collateral-based on SME lending: The role of business collateral and personal collateral in less-developed countries. *Research in International Business and Finance, 39*, 406–422. https://doi.org/10.1016/j.ribaf.2016.07.005

Essel, B. K. C., Adams, F., & Amankwah, K. (2019). Effect of entrepreneur, firm, and institutional characteristics on small-scale firm performance in Ghana. *Journal of Global Entrepreneurship Research, 9*(1), 1–20.

Ferri, G., Murro, P., Peruzzi, V., & Rotondi, Z. (2019). Bank lending technologies and credit availability in Europe: What can we learn from the crisis? *Journal of International Money Finance, 95*, 128–148.

Garson, G. D. (2012). *Testing statistical assumptions*. Statistical Associates Publishing.

Gehrig, T., & Stenbacka, R. (2007). Information sharing and lending market competition with switching costs and poaching. *European Economic Review, 51*(1), 77–99.

Hasanaj, P., & Kuqi, B. (2019). Analysis of financial statements: The importance of financial indicators in Enterprise. *Humanities and Social Science Research, 2*(2) 2019. ISSN 2576-3024; E-ISSN 2576-3032.

Hauswald, R., & Marquez, R. (2006). Competition and strategic information acquisition in credit markets. *Review of Financial Studies, 9*(3), 967–1000. https://doi.org/10.1093/rfs/hhj021

Kelava, A. (2016). A review of confirmatory factor analysis for applied research (2nd ed.). *Journal of Educational and Behavioral Statistics, 41*(4), 443–447. https://doi.org/10.3102/1076998616631747

Kessey, K. D. (2015). Assessing credit risk management practices in the banking industry of Ghana: Processes and challenges. *Global Journal of Management and Business Research, 15*, 1–10.

Li, Y., Lu, R., & Srinivasan, A. (2019). Relationship bank behavior during borrower distress. *Journal of Financial and Quantitative Analysis, 54*(3), 1231–1262. https://doi.org/10.1017/S0022109018001084

Manurung, E. (2019). A new approach of bank credit assessment for SMES. *Academy of Accounting and Financial Studies Journal, 23*, 22–35.

Mayo, H. (2012). *Principles of finance* (pp. 251–276). UET Press.

Mills, K., & McCarthy, B. (2014). *The state of small business lending: Credit access during the recovery and how technology may change the game* (Harvard Business School General Management Unit Working Paper [15-004]).

Na-Nan, K., & Wongsuwan, N. (2020). Development and validation of perceived workplace support for small and medium-sized enterprise employees. *International Journal of Organisational Analysis, 28*(1), 243–259. https://doi.org/10.1108/IJOA-12-2018-1608

Palmgren, P., Brodin, U., Nilsson, G., Watson, R., & Stenfors, T. (2018). Investigating psychometric properties and dimensional structure of an educational environment measure (DREEM) using Mokken scale analysis—A pragmatic approach. *BMC Medical Education, 18*, 235. https://doi.org/10.1186/s12909-018-1334-8

Petersen, M. A., & Rajan, R. G. (1994). The benefits of lending relationships: Evidence from small business data. *The Journal of Finance, 49*(1), 3–37.

Petersen, M. A., & Rajan, R. G. (1995). The effect of credit market competition on lending relationship. *Quarterly Journal of Economics, 110*, 407–443.

Poku, K., & Frimpong, F.A. (2009). *The assessment of the prospects of venture capital finance in Ghana: The views and perception of some selected SMEs in Ghana*. Ninth Annual IBER and TLC Conference Proceedings 2009, Las Vegas, NV, USA.

Price Waterhouse Coopers (PwC). (2019). *Ghana banking surveys*.

Quartey, P., Turkson, E., Abor, J. Y., & Iddrisua, A. M. (2017). Financing the growth of SMEs in Africa: What are the constraints to SME financing within ECOWAS? *Review of Development Finance, 7*, 18–28.

Ramalho, R., Jiang, N., Koltko, O., Chávez, É., Koch-Saldarriaga, K. A., & Quesada Gamez, M. A. (2018). *Improving access to finance for SMES: Opportunities through credit reporting, secured lending, and insolvency practices* (No. 129283, pp. 1–63). The World Bank.

Sette, E., & Gobbi, G. (2015). Relationship lending during a financial crisis. *Journal of the European Economic Association, 13*(3), 453–481.

Slocum-Gori, S. L., & Zumbo, B. D. (2011). Assessing the unidimensionality of psychological scales: Using multiple criteria from factor analysis. *Social Indicators Research, 102*(3), 443–461.

Soares, J. O., Pina, J., Ribeiro, M., & Lopes, M. C. (2011). Quantitative vs. qualitative criteria for credit risk assessment. *Frontiers in Finance and Economics, 8*(1), 69–87. https://doi.org/10.2139/ssrn.2012443

Tarka, P. (2018). An overview of structural equation modeling: Its beginnings, historical development, usefulness and controversies in the social sciences. *Quality & Quantity, 52*(1), 313–354. https://doi.org/10.1007/s11135-017-0469-8

Uchida, H. (2011). What do banks evaluate when they screen borrowers? Soft information, hard information, and collateral. *Journal of Financial Services Research, 40*(1), 29–48. https://doi.org/10.1007/s10693-010-0100-9

Win, S. (2018). What are the possible future research directions for bank's credit risk assessment research? A systematic review of literature. *International Economics and Economic Policy, 15*(4), 743–759. https://doi.org/10.1007/s10368-018-0412-z

World Bank. (2019). *Disruptive technologies in the credit information sharing industry: Developments and implications*. Fintec Note No. 3. https://doi.org/10.1596/31714

Yong, S., Kwenin, C., & Ingabo. C. (2021). *Financial constraints to entrepreneurship in Ghana*. A report developed as part of the Global Inclusive Growth Partnership (GIGP), in collaboration Aspen Institute and the Mastercard Center for Inclusive Growth.

Yoshino, N., & Taghizadeh-Hesary, F. (2018). *A comprehensive method for the credit risk assessment of small and medium-sized enterprises based on Asian data* (No. 907). ADBI Working Paper Series.

Index

J. A. Peprah et al. (eds.), *Financial Sector Development in Ghana*, Palgrave Macmillan Studies in Banking and Financial Institutions, https://doi.org/10.1007/978-3-031-09345-6

FSC
www.fsc.org
MIX
Papier | Fördert gute Waldnutzung
FSC® C083411

Kolibri 360 GmbH
Ferdinand-Jühlke-Straße 7
99095 Erfurt, Deutschland
produktsicherheit@kolibri360.de